Geopolitics and the Event

RGS-IBG Book Series

For further information about the series and a full list of published and forthcoming titles please visit www.rgsbookseries.com

Published

Geopolitics and the Event

Rethinking Britain's Iraq War Through Art

Alan Ingram

WILEY

Registered Offices
John Wiley & Sons, Inc., 111 River Street, Hoboken, NJ 07030, USA
John Wiley & Sons Ltd, The Atrium, Southern Gate, Chichester, West Sussex, PO19 8SQ, UK

Editorial Office
9600 Garsington Road, Oxford, OX4 2DQ, UK

For details of our global editorial offices, customer services, and more information about Wiley products visit us at www.wiley.com.

Wiley also publishes its books in a variety of electronic formats and by print-on-demand. Some content that appears in standard print versions of this book may not be available in other formats.

Library of Congress Cataloging-in-Publication Data applied for
9781119426004 (hardback); 9781119426059 (paperback)

Cover Design: Wiley
Cover Image: © kennardphillipps

Set in 10/12pt Plantin by SPi Global, Pondicherry, India
Printed and bound in Singapore by Markono Print Media Pte Ltd

10 9 8 7 6 5 4 3 2 1

Contents

List of Figures

Series Editor's Preface

The RGS-IBG Book Series only publishes work of the highest international standing. Its emphasis is on distinctive new developments in human and physical geography, although it is also open to contributions from cognate disciplines whose interests overlap with those of geographers. The Series places strong emphasis on theoretically-informed and empirically-strong texts. Reflecting the vibrant and diverse theoretical and empirical agendas that characterize the contemporary discipline, contributions are expected to inform, challenge and stimulate the reader. Overall, the RGS-IBG Book Series seeks to promote scholarly publications that leave an intellectual mark and change the way readers think about particular issues, methods or theories.

For details on how to submit a proposal please visit:
www.rgsbookseries.com

David Featherstone
University of Glasgow, UK
RGS-IBG Book Series Editor

Acknowledgements

My thanks to all interviewees and other participants in the research on which this book is based, who have been unfailingly generous in sharing their experiences, insights and work with me. Thanks also to the museum and gallery staff who have facilitated the research in many different ways, and to all the artists and institutions who have allowed images of artworks to be reproduced.

The project on which the book is based would not have been possible without a British Academy (BA) Mid-Career Fellowship, which allowed me a year to develop the research, and sabbatical leave from the Department of Geography at University College London (UCL), which enabled public engagement activities and the writing of the book itself. Particular thanks are due to Matthew Gandy for his encouragement and advice in the early stages of the project and book, and to María Rodríguez for her expert help in helping me apply to the BA. The project has also been supported by the UCL Public Engagement Unit, and I am particularly grateful to Laura Cream for her encouragement and advice in the early stages. A UCL Beacon Bursary award for public engagement enabled me to organise and curate *Geographies of War: Iraq Revisited* at UCL in March 2013 and I would like to thank Susie Chan, UCL Exhibitions Coordinator, who secured the space and logistical support, Cath D'Alton and Miles Irving from the UCL Geography Drawing Office, who designed and organised production of the catalogue, captions and publicity material, and the PhD and MSc students who invigilated and observed the exhibition. Particular thanks go to Peter Kennard, Cat Phillipps, Yousif Naser, Satta Hashem, Doug Farthing, Emily Johns and Hanaa Malallah for contributing to the exhibition, and to the Park Gallery for its loan of Hanaa Malallah's work *My Country Map* for its duration. I would also like to thank all speakers and participants in public engagement events around the exhibition, and the Ark in Acton and the Mosaic Rooms in Kensington for hosting events.

Over and above the research interviews conducted for the project, my knowledge and understanding has been enhanced greatly by conversations and correspondence with many people over the last 10 years, among whom I would particularly like to thank Rashad Selim, Robert Kluijver, Dan Gorman, Hassan Abdulrazzak, Nadje Al-Ali, Charles Tripp, Rijin Sahakian, Wafaa Bilal, War Boutique, Tim Shaw, Beau Beauregard, James Marriott, Michaela Crimmin, Mona Kriegler, Hydar Dewachi, Eugenie Dolberg, Martha Rosler, Susan Noyes Platt, Nicola Gauld, Ian Klinke, Andrew Barry, Jason Dittmer, Tariq Jazeel, Felix Ciuta and Matthew Shlomowitz. Thanks go especially to Felix and to Matthew for their reading, critique and discussion of various papers and chapters on art and geopolitics, and for ongoing conversations on art, music and popular culture. Thanks also to Matthew for inviting and enabling me to discuss my work with people in the contemporary classical music world, which has been challenging, enlightening, useful and enjoyable. Particular thanks also to participants in *Al-Mutanabbi Street Starts Here,* co-organised with Hassan Abdulrazzak at UCL in 2014, and to Eleanor Robson, Mina Al-Oraibi and Philippe Sands for their participation in *Thinking Through the Event: The Report of the Iraq Inquiry* at the UCL Institute of Advanced Studies in November 2016, both of which influenced my thinking in a number of ways. Taking part in two Martin Luther King Peace Days organised by Nick Megoran and colleagues at Newcastle challenged me to try to make the themes of my research more comprehensible and interesting to others and affected how I thought about the project.

The research on which the book is based has been presented at seminars and conferences at Cambridge, Oxford, UCL, Newcastle, Kings College London, Glasgow, Swansea, City, Northumbria, University College Cork, Amsterdam and Tampere; at the Royal Academy of Art, the Arts Catalyst, the Norwegian Academy of Music and the Accademia Europea di Firenze; and at annual international meetings of the Royal Geographical Society (RGS) and the Association of American Geographers. My thinking has been enriched by RGS-IBG conference sessions co-organised with Klaus Dodds and with Harriet Hawkins, and by many conversations with them and with Simon Dalby, Gerard Toal, Jo Sharp, Sara Koopman, Nick Megoran, Joe Painter, Stuart Elden, Gerry Kearns, Karen Till, Sara Fregonese and Ian Shaw, among many others, on geopolitics, nationalism, art, war and peace. I'm particularly grateful to Gerry, Nick, Simon and Joe and to Jenny Robinson for their support over the longer term.

My thinking has benefited greatly from the opportunity to work with excellent undergraduate, masters and doctoral students, as well as great colleagues, at UCL Department of Geography and from being able to maintain close links between teaching and research. My own research has been enhanced by acting as supervisor to Cinzia Polese, Ophélie Véron, Sam Halvorsen, Charlotte Whelan, Cat Gold and Lioba Hirsch, who have each been tremendously impressive graduate students. Prior to joining UCL I was lucky to have been able to study and work with many great academics, researchers and students at the Department of Geography and at St Catharine's College and Sidney Sussex College, Cambridge, among whom Linda McDowell and the late Graham

Smith, my PhD supervisor and, for a short time, colleague, have been particularly important. I am also hugely grateful to John Wyn Owen, Colin McInnes and Kelley Lee, with whom I worked closely during a short time out of academia, learning much in the process.

Many thanks to Dave Featherstone as editor of the RGS-IBG series for his encouragement and feedback, to the editorial board for their helpful comments, and to the reviewers of the proposal and manuscript for their careful, critical and constructive engagement, which enabled me to rethink and refine the book. Thank you also to Jacqueline Scott at Wiley for being such an excellent editor, to Elisha Benjamin for patient help with the often-complicated process of securing permission to reproduce images of artworks, to Rosemary Morlin, whose careful copy editing improved the manuscript, and to Shyamala Venkateswaran and colleagues for help in the final stages of production. Thanks to Peter Kennard and Cat Phillipps for permission to use a detail from the kennardphillipps work *Untitled (Iraq)* (2005) as the cover image. I remain responsible for everything contained herein.

Thank you to my parents Marlene and Alec and my sister Esther for their love and support, and to Clea, Maisie and Dylan for every day I get to share with them.

Chapters 2, 4 and 7 draw upon, expand and develop arguments and examples discussed in:
- Ingram, A. (2016). Rethinking art and geopolitics through aesthetics: artist responses to the Iraq war. *Transactions of the Institute of British Geographers* 41 (1): 1–13.

Chapter 3 draws on conceptual arguments developed in the following articles:
- Ingram, A. (2017). Art, geopolitics and metapolitics at Tate Galleries London. *Geopolitics* 22 (3): 719–739.
- Ingram, A., Forsyth, I. and Gauld, N. (2016). Beyond geopower: earthly and anthropic geopolitics in *The Great Game* by War Boutique. *cultural geographies* 23 (4): 635–652.

Chapter 4 draws upon and reinterprets an example discussed in:
- Ingram, A. (2012). Bringing war home: from Baghdad 5 March 2007 to London 9 September 2010. *Political Geography* 31: 61–63.

Chapter 6 draws upon and reinterprets examples discussed in:
- Ingram, A. (2009). Art and the geopolitical: remapping security at *Green Zone/ Red Zone*. In: *Spaces of Security and Insecurity: Geographies of the War on Terror.* (eds. A. Ingram and K. Dodds), 257–277. Farnham: Ashgate.

Chapter One
Introduction

Laid out during the middle of the nineteenth century, London's Trafalgar Square embodies and displays geopolitical power. At its centre is a massive stone column supporting a statue of Admiral Horatio Nelson, to many, Britain's foremost naval commander. Overlooking the Square from the north side is the National Gallery, an institution that contains one of the greatest collections of European art, while on other sides are located the High Commissions of former British imperial territories that are now independent states. A short walk away down Whitehall lie the main offices of government and state. In close proximity are many other leading cultural institutions, including the National Portrait Gallery, Tate Britain and Tate Modern, each of which has historically tended to focus on European and North American art. The British Museum, which holds one of the world's foremost collections of ancient artefacts, is located a short distance away.[1] At the same time, the Square has also always invited appropriation and subversion. Constructed to assert the grandeur and authority of the British state, military and empire at a time of wars and revolutions, and often hosting officially-endorsed events of national significance, the Square has periodically been taken over by protests and demonstrations, including, in the recent past, protests against British participation in the wars in Afghanistan and Iraq.

At three corners of the Square are located plinths supporting statues of military and political leaders, but the fourth plinth, in the north west corner, was left vacant after a subscription campaign failed to raise sufficient funds to pay for a statue to go there. Since 1998, this 'Fourth Plinth' has hosted a series of installations by

Geopolitics and the Event: Rethinking Britain's Iraq War Through Art, First Edition. Alan Ingram.
© 2019 Royal Geographical Society (with the Institute of British Geographers).
Published 2019 by John Wiley & Sons Ltd.

prominent contemporary artists, who are periodically selected by commission and whose proposed work is then installed there for a few months at a time (de Vasconcellos 2016). The Plinth has offered selected artists an opportunity to introduce something different into public space, something that might be more, less or other than a protest, and which might engage the public and enter into conversation with its surroundings (Sumartojo 2013).

In 2008, the British artist Jeremy Deller was among those invited to propose a work. Troubled by what he saw as the restricted ways in which the ongoing war in Iraq was being represented and debated, Deller proposed that an object such as a damaged car should be relocated from Iraq and exhibited on the Plinth with the title *The Spoils of War*, a harshly ironic reference to the imperial practice of displaying objects taken from conquered lands. The appearance of an actual object from Iraq, destroyed in a conflict in which British forces continued to be engaged, might, the artist hoped, disrupt public discourse about the war. While Deller constructed a photomontage to illustrate his idea (Figure 1.1) and a maquette for the work was displayed for a few weeks along with other proposals in a small anteroom to the National Gallery, his idea was not adopted having been judged unsuitable, or at least not the most suitable. Deller did subsequently succeed in

Figure 1.1 *The Spoils of War*, Jeremy Deller (2008). Image courtesy of the artist.

having a wrecked car from Iraq exhibited in the Imperial War Museum (IWM) in London, but only after British forces had officially withdrawn from the country.[2]

In late 2017, nearly ten years on and several commissions later, it was announced that a work in the series *The Invisible Enemy Should Not Exist* by Michael Rakowitz (an American artist of Iraqi Jewish heritage) had been selected to appear on the Plinth. The proposed object would be a reproduction of a *lamassu*, a protective deity in the hybrid form of a winged bull with a human head. The original had stood at the gates of the Assyrian city of Nineveh in northern Iraq, from around 700 BCE until it was destroyed by Islamic State (IS) militants after they took over the area in 2015 (Greater London Authority 2018).[3]

The *lamassu* (Figure 1.2) was unveiled in March 2018, and the work can readily be interpreted as an act of resistance and affirmation in the face of the

Figure 1.2   *The Invisible Enemy Should Not Exist,* Michael Rakowitz (2018). A Mayor of London Fourth Plinth Commission. Photo by author, courtesy of the artist.

brutal iconoclasm of IS. The sculpture also evoked the 2003 invasion and subsequent occupation of Iraq by the United States and Britain, which created the conditions in which a movement like IS could emerge and thrive, and, more specifically, the fact that, while they had defeated the Iraqi military and occupied the country's cities in a matter of weeks, American and British forces had failed to protect Iraq's museums, galleries, archives and libraries, which were subject to extensive looting, vandalism and destruction (Bahrani 2003; see also Bogdanos 2005). The sculpture would serve as a reproduction of a lost object, while bearing a complex relation to the original and carrying wider symbolic and material resonances. As Rakowitz stated, 'I see this work as a ghost of the original, and as a placeholder for those human lives that cannot be reconstructed, that are still searching for sanctuary' (Elbaor 2017). In emphasising its ghostliness, Rakowitz evoked questions of spectrality, of what can be seen and what cannot, of appearance and liminality (Pilar Blanco and Peeren 2013, p.1). Like Deller's proposal for the wrecked car, the *lamassu* conjures up bodies both corporeal and geopolitical, staging a haunting return for people and objects that have been lost through war.

As with other objects in the *Invisible Enemy* series, the *lamassu* was fabricated out of materials used to package and sell groceries produced in Iraq and exported beyond the country, in this case date syrup cans. Through the use of everyday objects, the work would, in the words of the artist, further embody Iraq's 'former economic power, now destroyed by war' (Elbaor 2017), as well as the conditions of displacement, exile and diaspora experienced by millions of Iraqi people. The location of the work was also highly significant. While previous exhibitions of works in the series had taken place in museums and galleries, here the sculpture appeared not just in a major public space, but at a site of huge material and symbolic importance to Britain's military and colonial past, and to its ongoing political and cultural life. On the Fourth Plinth, the *lamassu* provided a dramatic contrast with the other statues in the Square, while seeming to form a strange alliance with several original *lamassu* taken from Iraq by nineteenth-century colonial archaeologists and which continue to be displayed in the British Museum.

The appearance of the *lamassu* was an event in its own right, but one implicated in many other events, which it might be said to have embodied and activated in a variety of ways. Its appearance was especially significant in a situation where, as with so much of the country's colonial history, many of Britain's political and cultural institutions have struggled, or simply failed to recognise or comprehend, the scope, nature and implications of the 2003 Iraq war. While several major public inquiries into aspects of the war have made sporadic reference to the harms suffered by Iraqi people, they have focused principally on Britain, being concerned primarily with British politicians, civil servants, spies, generals, soldiers and administrators; with what they did, what they did wrong, what they did not do, and what they might have done differently. Iraq and Iraqi people, for whom the war was in many ways a quite different kind of event, hardly appear. The *lamassu*, by contrast, offers a different way into the event. As this book argues, artworks like the *lamassu* expand and complicate what is often taken to be the event of the 2003 Iraq war and point towards the multiple nature, or multiplicity, of geopolitical events more generally.

Official inquiries into the 2003 war have not been tasked with investigating the effects of British colonial influence on Iraq, or of recurring intervention in the country, but such exclusions reflect nonetheless a broader situation in which the appearance and participation of 'other' people and places in British public life are heavily conditioned and circumscribed. Jacques Rancière (2006) conceptualises this process of determining what and whom can appear and be recognised in the public sphere as *le partage du sensible*: the distribution, sharing or division of that which is sensible, or available to sense perception (see also Dikeç 2005; 2013; Dixon 2009; Dixon 2015). Rancière (2006) further argues that the issue of who and what can appear and be seen, and of who can speak and be heard in public, is prior to most definitions of politics. The 'distribution of the sensible', he argues, represents a 'primary aesthetics' that conditions the political realm and the ways in which events can appear as matters for public experience, deliberation and debate (Rancière 2006, pp.12–13). This idea can be linked with the argument advanced by Edward Said (1978) that Europeans have, over an extended period, represented the people and places of the Middle East in ways that systematically diminish their agency, diversity and voice, and ultimately their humanity, a phenomenon he called Orientalism. To the extent that people and places subject to Orientalism appear in Western public life, they do so in a limited and prejudicial manner. As Gregory (2004a, p.253) writes, building on Said, 'colonial modernity is intrinsically *territorializing*, forever installing partitions between "them" and "us"'. One consequence of this is to inhibit the extent to which events might be encountered and apprehended as taking place within a common world.

The *lammasu* brought what were described in its caption as 'the Iraq wars' into public space in Britain in a new and arresting manner, complicating existing distributions and configurations, and appearing to pose the possibility of a common world linking the two countries. This would, however, be a curiously limited intervention, which people would be free to ignore, and its ghostly presence would only appear for a fixed period of time, while the three other plinths and Nelson himself would remain in place long after it had departed. However, as this book explores, we can also approach the Iraq wars, and the 2003 war in particular, via dozens upon dozens of other artworks and exhibitions that have taken place in Britain. The book is based on a study of many of these works and exhibitions, which, it argues, both express and reassemble Britain's Iraq war, resisting, challenging and moving beyond the kinds of accounts offered by broadcast media, politicians, journalists, soldiers, lawyers and campaigners and by official inquiries. As the book shows, the 2003 Iraq war has given rise to numerous examples of critical, creative and imaginative works that in many cases can be said to oppose the war, but which all concentrate attention on how the event of war is conducted and experienced aesthetically: on how it is embodied, how it is sensed and made sense of, on how it resists sense-making and fails to make sense, and on how war both constitutes, and is brought into, the public sphere. Because such processes are intrinsic to how people experience the world and conduct politics in it, the book argues, an exploration of works like the wrecked car and the *lamassu* offers distinct insights into what we understand geopolitical events to be.

The 2003 invasion and subsequent occupation of Iraq by the United States, Britain and their allies led to the deaths of hundreds of thousands of people, the displacement of millions, the collapse of public order and the emergence of new forms of authoritarianism and extremism. Urban infrastructures and public services were devastated and much of Iraq's cultural heritage was destroyed, damaged or stolen. The 'Iraq war', as it is commonly called, is widely regarded as one of the greatest foreign policy disasters of modern times and a major violation of international law. Its consequences continue to reverberate throughout the country, the region and beyond.[4]

While the political, legal, diplomatic and military dimensions of the war have been the subject of numerous academic, journalistic and judicial inquiries, this book offers a new perspective by exploring how this event has been encountered, appropriated and reworked in art.[5] The invasion and occupation caused severe damage to the cultural heritage and cultural life of Iraq, but they also prompted the creation of many artworks within Iraq and beyond, which form alternative expressions and accounts of the event out of which they have emerged. Focusing critically on Britain as a central actor in the war and the former colonial power in Iraq, the book looks at the diverse ways in which artists experienced this event, the works they created, and the ways in which these works have been curated, exhibited or otherwise brought into the public sphere in Britain, thereby taking part in broader struggles and debates while often retaining a strange distance from them.[6] This provides a complement to other accounts of the war, showing how it was played out in the cultural as well as political sphere, but the broader argument of the book is that artistic enactments of the war not only challenge a straightforward understanding of it as 'an' event, but push us to rethink the nature of geopolitical events more generally, as multiple as well as singular things.

The book argues that a consideration of art has much to contribute to the genealogical analysis of geopolitics, that is, to a form of analysis that seeks to question the apparent self-evidence of events and to consider the possibility of understanding and enacting them in different ways. Borrowing an idea from Gilles Deleuze (2015) and building on the work of Jill Bennett (2012), the book interprets artworks in terms of the 'counter-actualisation' of the event, a process in which artworks can be understood as appropriating events and enacting them otherwise. In so doing, they allow people to enter into and go through the event in ways that differ from other ways of encountering it, thereby offering a counterpoint to ordinary experience. As the book argues, approaching the Iraq war in terms of its counter-actualisation in art reveals a field of creative resistance to the event, but also enhances our ability to understand it – and by extension other geopolitical events – as being both composed of, and situated in relation to, many other events, as well as being experienced and enacted in often radically different ways by different people, and, while not absolving people of their specific responsibility for the event, as taking place beyond the grasp of any one individual or group.

In pursuing a genealogical analysis, the book builds on other studies of the Iraq war and of contemporary geopolitics, but seeks to go beyond them in a number of ways. One key reference point is Derek Gregory's *The Colonial Present*, a book that was completed in the early months of the occupation and which ana-lysed the resurgence of colonial imaginaries in the context of the war on terror. As that book concludes,

> for us to cease turning on the treadmill of the colonial present … it will be necessary to explore other spatializations and other topologies, and to turn our imaginative geographies into geographical imaginations that can enlarge and enhance our sense of the world and enable us to situate ourselves within it with care, concern and humility. (Gregory 2004a, p.262)

Over the course of the last decade or so, academics working in political geog-raphy, international relations (IR) and related fields have indeed paid growing attention to the alternative spatialisations, topologies and imaginaries enacted by artworks engaging with war, militarism and security (e.g. Bleiker 2009; Graham 2010; Danchev 2011; Shapiro 2012; Amoore 2013). Their work can be related to a broader interdisciplinary movement in which art is seen as being not just reflective of historical, political and geographical developments, or as subor-dinate to, or illustrative of, conceptual inquiry, but as a productive field of exper-imentation and critique in its own right (e.g. van Alphen 2005; Jill Bennett 2005, 2012; Thompson and Independent Curators International 2009; Dear *et al.* 2011). Where critical discussion has touched upon art related to the 2003 Iraq war, however, this has often been fleeting (e.g. Sylvester 2009; Berlant 2011), and has, beyond isolated examples (e.g. Gregory 2010a; Platt 2011), largely failed to engage with the work of artists, curators and academics of Iraqi heritage within and beyond Iraq, many of whom have come to be based in Britain, have worked here, or whose work bears on Britain's role in relation to Iraq. A fuller consideration of their work underscores the multiplicity of the 2003 Iraq war and poses the question of how we might, via an appreciation of this multiplicity, still understand it as being 'an' event.

A deeper issue with existing literature in political geography and international relations is that it has, with few exceptions (e.g. Bleiker 2009; Sylvester 2009; Danchev 2011), tended not to think very genealogically about art, at least in the sense that I develop here. At one level, there has sometimes been limited art historical contextualisation of particular artworks or art forms, and limited recognition of their ontological complexity. As a result, artworks have sometimes been thought of as resisting geopolitical power (or criticised for failing to resist it) in an overly straightforward way. Still more fundamentally, there has been insufficient recognition of the extent to which the field of art has itself been implicated and invested in the forms of geopolitical power that artworks may be held to critique, resist and move beyond. As the book argues, it is necessary to recognise more fully that the idea of art as a singular, quasi-autonomous form of practice and experience that is typically at work in critical accounts – and which nationalist and anti-colonial movements have often tried to mobilise in the service of self-determination – emerged in the context of the modern Western European nation state and the forms of capitalism, patriarchy and colonialism associated with it, and that, while it may offer critical potential, this form of art has also long been used to assert, define and justify the superior status of some kinds of people over others (Stoler 2008). Any appeal to the critical potential of art must therefore be alive to its ongoing as well as historical entanglement with geopolitical power.

The book is centrally concerned with what artworks are and what they do, with how they get made and with how they come to acquire public existence, reflecting, participating in and contributing to broader struggles and debates, while not being reducible to them. In order to think about how art can be said to counter-actualise geopolitical events, the book conceptualises art works as evental assemblages, or heterogeneous constellations of things that emerge in relation to events, which are themselves events, and which may engender and participate in further events. If artworks are able to counter-actualise events, it is because they are themselves particular kinds of events.

The approach and arguments of the book can further be located in relation to three broad streams of thought in which ideas of art, aesthetics and geopolitics are being reworked. These streams of thought inform the book as a whole and serve as important reference points in the chapters that follow.

The first line of thinking emerges from what has come to be known as the decolonial project, a set of epistemic and political interventions that challenge what Mignolo (2009, p.162) calls the 'colonial matrix of power'. Seeking to go beyond perceived limits in postcolonial critique and practice, the decolonial project aims to enable the production of knowledge against and beyond this matrix, and to engage in the direct decolonisation of institutions and relationships (see for example Esson *et al.* 2017). This project has fundamental implications for ideas of geopolitics on the one hand, and art, aesthetics and the humanities on the other, the dominant meaning of each having emerged out of modern European nation-, state- and empire-building. It also has implications for how we think about events.

Approaching the Iraq war in terms of its coloniality requires an acknowledgement not just of how Britain, as a colonial power, 'made' Iraq, but of how 'Britain' has in some ways itself been constituted through its colonisation of Iraq, and of how this process has been aesthetic as well as political. This process becomes apparent if one is able to visit the British Museum, an institution founded in the nineteenth century upon the use of artefacts taken from other countries and displayed in order to foster not just understanding of the world, but a sense of British scientific, aesthetic and racial primacy.[7] The Museum's Mesopotamian Galleries display many beautiful and impressive objects taken from the region, including the *lamassu* mentioned earlier, which were relegated by nineteenth-century museum managers from the category of 'art' to that of 'artefact' (Bohrer 1998; Bahrani 2001; Sylvester 2009). While this distinction has often been challenged in recent museological, curatorial and artistic practices, the deeply sedimented material, institutional, discursive and aesthetic traces of colonialism and contemporary manifestations of coloniality continue to shape how 'Iraq' can appear in Britain. Approaching *Britain's* Iraq war through art takes us beyond invocations of coloniality as condition, towards a critique of a specific geopolitical culture (Toal 2008) and the associated institutions, discourses, sites and practices through which Iraq has been constituted as a concern and problem for Britain.

Such considerations indicate a degree of alignment between genealogical analyses of geopolitics, which question singular accounts of nation, state and empire, and the decolonial project, but this book cannot be said to engage directly in decolonial work. As Vazquez and Mignolo (2013, unpaginated) state, 'The decolonial option operates from the margins and beyond the margins of the modern/colonial'. This is not where I am coming from. As a white, English-speaking, British academic who has grown up, lived and worked in Britain, has never visited Iraq, does not speak Arabic and has not experienced war directly, I am not working from the margins or from beyond them, but from a geopolitical culture that is centrally implicated in the event. In this book, however, I do not aim or claim to represent Iraq or Iraqi experience directly, or in an ethnographic or sociological way. Rather, in much of what follows, I have taken cues from the extensive body of work through which artists, curators, academics and writers of Iraqi heritage have already framed and presented their experiences, thought and creative practice, in English.[8] I do not claim that the book enacts a decolonial intervention, but examine coloniality and a specific geopolitical culture in terms of how they have been questioned, resisted and reworked through art, and consider what the limits to this process might be.

The book is also written in light of feminist scholarship on geopolitics, war and peace, which has been of fundamental importance in opening up ways of thinking about power and aesthetics beyond the state. Drawing in part on feminist approaches to international relations, early feminist scholarship on geopolitics examined the gendered nature of geopolitical discourse and highlighted the body as a site through which geopolitical power was exercised, and in relation to which it could be critiqued (Dowler and Sharp 2001; Hyndman 2004). Feminist approaches to geopolitics have developed and diversified in a number of ways, including recognising the intersectional nature of patriarchal and colonial power (Dixon and Marston 2013; Massaro and Williams 2013). Particularly relevant to the approach and arguments of this book is a continued questioning not just of the epistemology and practice of geopolitics but of its ontology, in particular its framing of state and territory as central categories of concern, seeing these rather as being 'enrolled' in particular kinds of practices and materialities not conventionally thought to be geopolitical (Dixon 2015; also Dixon *et al.* 2012; Yusoff 2015). In attending to the transnational dimensions of geopolitical events, the book also chimes with moves in feminist IR arguing for a reorientation in the study of war away from the state, and towards 'a set of experiences that everyday people [as well as] elites have emotionally, physically, and socio-ethically, depending on their locations inside and even far beyond war zones' (Sylvester 2013, p.65, emphasis added; also, for example, Hörschelmann 2008; Hörschelmann and Refaie 2014). In focusing on art, the book's concerns also chime with feminist work (e.g. Dowler and Sharp 2001; Koopman 2011) exploring how geopolitics might be enacted differently. To develop its arguments the book also draws on feminist critiques of aesthetics (e.g. Armstrong 2000) and feminist theorisations of art (e.g. Pollock 2013a, 2013b). To the extent that feminist work and feminist geopolitics in particular are premised upon certain kinds of embodiment and experience, however, the book, again, does not claim to enact a feminist intervention but rather works in alignment and conversation with certain strands of feminist argument and analysis.

The third stream of critical thought informing the book concerns the idea of the anthropocene as a new geological epoch in which humans, or at least certain groups of humans, have come to act as a fundamental influence on earth system processes, necessitating a fundamental rethinking of the meaning of geopolitics, of the human and also of the event (e.g. Dalby 2007; Clark 2014; Dixon 2015; Yusoff 2016; Bonneuil and Fressoz 2017; Davis and Todd 2017). In light of these developments, and, prompted especially by the many artworks emerging from the war that deal with matters of oil, I approach the 2003 Iraq war as an *anthropocenic* war in the following ways. As has been widely discussed and debated, the war can be regarded as a war 'for' oil, coming at a peak in concerns about 'peak oil' and American hegemony after the end of the Cold War (Harvey 2003; Gregory 2004a). As a geopolitical enterprise, the invasion and occupation – as well as the 'war on terror' more widely – themselves consumed vast quantities of petroleum and petroleum-based resources (Bonneuil and Fressoz 2017): this has been a war *of* oil as well as a war *for* oil. Furthermore, in invading and occupying Iraq, US and British forces went into a country that not only forms one of the heartlands of the modern oil industry but a region where oil has been at the centre of social, cultural and political practices for centuries, if not millennia. Finally, since 2003 the role of oil companies in supporting arts institutions in Britain has increasingly been politicised, with the Tate ending its relationship with one of its major donors as a result, and the British Museum and other institutions coming under increasing pressure to do the same (Miller 2015; Ingram 2017). While many artworks emerging from the war have involved oil and the forms of geopolitics associated with it, the institutions within and around which they have often emerged should also be understood as being implicated in the production of the anthropocene as contemporary condition and problem.

The book advances three main arguments. It argues, first, for approaching Britain's Iraq war through art as a way of rethinking one of the most contentious and consequential geopolitical events of recent decades, and for doing so via a sustained consideration of diverse artworks and exhibitions, over an extended period of time. This provides new, critical perspective on an event that continues to affect political life in Britain as well as Iraq, and with which British public and political culture struggles to reckon. This implies a second, broader argument, that an engagement with art is instructive for the analysis of geopolitical events, but the book argues more specifically that a rethinking of how exactly art and geopolitics might be said to enact each other is required. The book develops this argument by thinking about art as a *dispositif* (Foucault 2000; Rancière 2009a) and by conceptualising artworks as evental assemblages, or as particular kinds of events that are bound up with geopolitics in complex, evental, ways via the process of counter-actualisation. The third, broadest, argument of the book is for developing the event as a central category for critical work on geopolitics, in that this can extend our thinking on some of the fundamental questions of ontology and epistemology with which the field is concerned.

The following chapters develop and exemplify these arguments, themes and concerns in different ways. Prompted by my own experience of, and research on, artworks, exhibitions and other events through which the war has been engaged and addressed in Britain and beyond, they trace diverse trajectories and pathways through the event and seek to question, enrich and expand the ways in which it might be understood and encountered, while still being somehow comprehensible as 'an' event.

Chapter Two works through the problem of the event in light of longstanding interest in this concept across the social sciences and humanities and recent discussions in political geography. Considering the often troubling, paradoxical and confounding nature of events, and echoing other recent interventions (e.g. Clark 2014; Bonneuil and Fressoz 2017), it conceptualises a geopolitical event as a disruptive transformation of the world and of the ability to sense and make sense of it. Engaging with the influential accounts of the event offered by a select group of French philosophers since the Second World War, the chapter also makes connections with feminist, postcolonial and decolonial thought and with critical work on the anthropocene in order to develop, through a series of propositions, a sense of geopolitical events as multiple as well as singular things, and a consideration of how, in the face of their violence and their confounding nature, geopolitical events might be approached and appropriated in a critical manner. The discussion begins to formulate the idea of Britain's Iraq war as a complex, multiple event and points towards the ways in which artworks might be said to counter-actualise events.

Chapter Three critiques existing literature on art and geopolitics and develops the idea of artworks as evental assemblages, or heterogeneous constellations of elements that emerge from events, which themselves constitute events and which are capable of eliciting and engendering further events. In order to think more about how it is that artworks are able to do this, the chapter develops Rancière's (2009a) idea of art as a *dispositif*. It works through debates on affect, posthumanism and sense-making, and turns to Raymond Williams' (1997) idea of structures of feeling as simultaneously affective, semantic, and embedded in practices. It then works Rancière's account of the *dispositif* of art through decolonial, Marxist and feminist critiques of aesthetics before reconsidering the Deleuzian idea of counter-actualisation, building on the work of Jill Bennett (2005, 2012) to argue for tracking the counter-actualisation of an ostensible event across multiple artworks, communities of sense and structures of feeling.

Chapter Four begins to map artistic enactments of the war in Britain, tracing the diverse pathways whereby artworks and exhibitions emerged in relation to the shifting politics of the war and exploring the important roles played by the British Museum, Institute of Contemporary Arts, Imperial War Museum and National Army Museum in mediating this process. The chapter develops the idea of museums as geopolitical institutions and explores how the distinction between art and artefact that is enacted by museums, a distinction in which objects extracted from Iraq have played a central role, comes to constitute a geopolitical division of the world. Building on postcolonial and decolonial critiques, it shows how artistic and curatorial enactments of the war have been both enabled and framed by museum institutions, demonstrating how the ability of artworks to counter-actualise geopolitical events is contingent upon a *dispositif* of art that is itself implicated in the ongoing breaking and making of the world.

While artworks and exhibitions created by artists and curators of Iraqi heritage within and beyond Iraq are discussed throughout the book, Chapter Five focuses on several artists of Iraqi heritage and artworks created by them relating directly to the war. While approaching artworks in terms of their political contexts and framings risks becoming reductive, the chapter notes how questions of national identity, autonomy, freedom and political conflict have, along with the aesthetics of ancient civilisations, formed an important and enduring reference point for many Iraqi artists since the emergence of modern Iraqi art in the middle of the twentieth century, and how these questions have acquired further layers with the development of diverse Iraqi diaspora communities. After considering the relationship of Iraqi art to modernism, the chapter focuses on the work of artists embodying distinct personal, aesthetic and political trajectories and concerns. The chapter both reconnects with themes raised by Rakowitz's *lamassu* and demonstrates the diverse ways in which the event of war has been encountered, appropriated and enacted, further illustrating how an engagement with art contributes to a sense of geopolitical events as multiple as well as singular things and how events may be counter-actualised in artworks.[9]

Chapter Six considers how the materialities and politics of oil have been appropriated and mobilised in artworks responding to the war in relation to questions of the anthropocene, coloniality and embodiment. After a brief survey of oil-related artworks, the chapter centres on a discussion of three sculptural installations that use oil or oil-related materials to configure an exhibition space as a whole: *Souvenir from the Ministry of Justice* by Rashad Selim, *20:50* by Richard Wilson and *Black Smoke Rising* by Tim Shaw. While *20:50*, a work first created in London, gained new geopolitical resonances as a result of its installation in a former regime prison in Iraqi Kurdistan, *Souvenir...* and *Black Smoke Rising* address oil-related geopolitical violence within the works themselves. The chapter argues that the use of oil in these works not only resonates with the idea of the 2003 war as a war for and of oil, but suggests an idea of art as an event that, like geopolitical violence and ultimately the anthropocene itself, places the human body and its relation to the world profoundly in question.

Chapter Seven considers how the 2003 war has been engaged through a specific art form, photomontage, and provides an account of how one photomontage in particular, kennardphillipps' *Photo Op*, has come to function as a definitive artwork of Britain's Iraq war. The chapter first rethinks photomontage via a critique of the reception of the photomontage work of the American artist Martha Rosler in writing on geopolitics, drawing out the multiple ways in which photomontage is entangled with events. It then explores the work of kennardphillipps and *Photo Op* in particular in relation to this revised account, exploring the complex implication of photomontages in the ongoing events of which they are counter-actualisations.

Chapter Eight draws on feminist and feminist-materialist thinking and reconnects with themes of embodiment, coloniality and the anthropocene to consider further how the figure of the body has appeared in artistic responses to the war and how those responses push our understandings of what a geopolitical body might be. The chapter considers how a series of artworks and exhibitions have registered and addressed the violence of the Iraq war, and explores the complicated ethical and political questions they suggest, through and beyond human embodiment. The later parts of the chapter consider how artworks created by people of Iraqi heritage have started to explore ways of living with, through and potentially beyond disaster via a variety of human and more-than-human bodies, and, reaching beyond Britain, the ways in which the Iraq National Pavilion at the Venice Biennale has come to form an important but also contested site at which the event of the war continues to be appropriated and explored. In considering debates surrounding the Pavilion, the chapter returns to consider how the capacity of artworks to operate as evental assemblages remains contingent upon a *dispositif* of art that continues to be implicated in geopolitical power.

The concluding chapter revisits the arguments of the book and considers some of the implications of the analysis. The book ends by revisiting the *lamassu* in Trafalgar Square.

Notes

1   Whitehead *et al.* (2014) open their introduction to political geography with a vignette that also considers Trafalgar Square as a site of geopolitical power.
2   Deller has explained his thinking on the car project on many occasions, including at discussion events at the National Gallery in early 2008, at IWM London in September 2010, and at the Hayward Gallery in October 2012. The displaying of the car at IWM London is discussed in Chapter 4.
3   In 2016, a scaled-down, 3D printed replica of an ancient arch destroyed by ISIS in Syria the previous year was installed in the Square, triggering critiques and debates about the politics of such acts (Burch 2017).
4   The Costs of War project based at Brown University in the United States has since 2011 aimed to assess 'the costs of the post-9/11 wars in Iraq and Afghanistan, and the related violence in Pakistan and Syria' (Costs of War 2018). In March 2013 it assessed recorded violent civilian deaths in Iraq since 2003 at 134,000, but noted that the actual figure could be double this (Crawford 2013). In April 2018, following the rise and apparent fall of the Islamic State (IS) movement, the Iraq Body Count project had assessed documented violent civilian deaths in Iraq since 2003 to be between 181,113 and 203,136; when combatants were included the estimate came to 288,000 (Iraq Body Count 2018). The number of people seriously injured is typically estimated as being a similar number to those killed; writing for the Costs of War, Crawford (2013, p.1) also notes that, 'many times the number killed by direct violence have likely died due to the effects of the destruction of Iraq's infrastructure'. I discuss the politics of body counts further in Chapter 4.
5   A brief list would include, among academic works: Harvey (2003); Gregory (2004a); Boal *et al.* (2005); Herring and Rangwala (2006); Fawn and Hinnebusch (2006); Al-Ali and Pratt (2009); Rappert (2012); Dodge (2012); Bailey *et al.* (2014); journalistic analysis: Keegan (2005); Chandrasekaran (2007); Cockburn (2006); Ricks (2006); Steele (2009); Fairweather (2011); memoirs and personal accounts: Pax (2003); Riverbend (2006); Stewart (2006); Allawi (2007); activist/policy analysis: Ledwidge (2011); Muttit (2011). The report of the Iraq Inquiry, the third major inquiry connected with the war commissioned by the British government, was published on 6 July 2016. See http://webarchive.nationalarchives.gov.uk/20171123122743/http://www.iraqinquiry.org.uk/the-report/ (accessed 3 April 2018).
6   The book does not aim or claim to represent the effect of the war on people or cultural life in Iraq directly, but it does consider how artists and others working and living in Britain and beyond have addressed these issues via their own experiences and practices.
7   As I explore in Chapter 4, the emergence of extractive colonial interests in Iraq's petroleum reserves in the twentieth century was foreshadowed by the development of extractive archaeological and museological practices in the nineteenth.
8   In conducting interviews for the project I aimed to be open about my positionality and modest about my experience, inviting participants whose work was already in the public domain to help me understand their experience and work. Participants expressed informed consent and have had the opportunity to review how they have been quoted and how their work has been discussed, in some cases suggesting amendments. I do

not claim that this can or does circumvent issues of positionality and power in research, however. The territorial nation state, the art world and the invitation to participate in academic research each induce, invite and sometimes compel people to present themselves and their work in certain ways. While seeking to mitigate this and to work with respect, fairness and accuracy, academic analysis and writing necessarily involves the exercise of certain kinds of analytical sovereignty and privilege, for which I take responsibility.

9   The term 'ancient' is loaded; in this context it by no means implies primitive; as Zainab Bahrani's (2001, 2008, 2014) work has demonstrated, these civilisations developed sophisticated aesthetic forms.

Chapter Two
Thinking Geopolitics Through the Event

An event is typically understood as something that happens. As the Oxford English Dictionary (2018, II.3a) tells us, establishing right away that events are spatial as well as temporal, an event is 'Something that happens or takes place'. An event is an ostensibly singular thing, and therefore nameable as such, which stands out from the background of normality and seems to make a difference in the world (Devetak 2009). As research in the cognitive sciences (Tversky and Zacks 2013; Radvansky and Zacks 2014) has begun to explore, an ability to recognise and make sense of events, as well as to initiate and shape them, would appear to be a fundamental, if not definitive, capacity of human beings. As Radvansky and Zacks (2014, p.38) summarise, events 'are a fundamental aspect of cognition [that] give our thoughts and action purpose and are the basis of our intelligent understanding of the world'. While this is an intriguing place from which to begin, scholars working on geopolitics are usually interested in events many times more complex than those typically encountered in psychology laboratories, from mundane occurrences that might or might not amount to a security concern (Anderson and Gordon 2016) to the geological scales of the anthropocene (Lewis and Maslin 2015).

While recognising what would seem to be the evental capacities of human beings, I work here with the genealogical understanding of the event that has emerged across the humanities and social sciences in recent decades (in political geography see for example Ó Tuathail 1996; Clark 2014; Yusoff 2016). The central feature of the event as it is understood genealogically is its indeterminacy

Geopolitics and the Event: Rethinking Britain's Iraq War Through Art, First Edition. Alan Ingram.
© 2019 Royal Geographical Society (with the Institute of British Geographers).
Published 2019 by John Wiley & Sons Ltd.

(Foucault 2000; Colwell 1997; James Williams 2009; Lundborg 2009). From a genealogical perspective, it is not possible to specify with precision or finality what any particular event (including its name and spatio-temporal parameters) is, partly because such designations are understood as being (in however small a way) part of the event itself and as thereby contributing to its ongoing actualisation. A genealogical conception of the event allows us to think beyond an event (such as 'the Iraq war' or 'Britain's Iraq war') as it ostensibly appears within a given frame of reference, to explore it in terms of its indeterminacy and its multiplicity. I suggest that it is only via a consideration of the multiplicity of a geopolitical event that it can properly be understood as 'an' event at all.

In this chapter, I conceptualise a geopolitical event as a disruptive transformation of the world and of ways of sensing and making sense of it. This conceptualisation emerges out of an engagement with a number of intersecting literatures concerning the event and certain events in particular, and it echoes in key respects ideas recently deployed by political geographers and others examining a range of shocking, contentious and transformative occurrences (especially, Kaiser 2012; Klinke 2013; Clark 2014; Bassett 2016; Yusoff 2016; Bonneuil and Fressoz 2017; also Adey *et al.* 2011; Shaw 2012; Toal and O'Loughlin 2017). This conceptualisation recognises, first, that events make a difference to things in that they enact (while not being identical or reducible to) intensive material, energetic and semiotic processes (DeLanda 1997), in the course of which things affect and alter each other, and appear and disappear. Second, this conceptualisation emphasises that a *geopolitical* event involves some kind of change in the world (also Shaw 2012), or part or parts of it, entailing such things as rupture, division, unification, invasion, occupation, emancipation, elimination, destruction, rearrangement or reconstruction. Geopolitical events are episodes in which the world – or, at least, a world – is broken and remade.[1] Understood in this way, an event is distinct from a process or a relation; it is, rather, something that marks a break, shift or bifurcation in the way things are and the ways they work. Third, a geopolitical event is in some way a forceful embodied occurrence, in that it challenges established ways of being and feeling in the world, in ways that can be difficult to perceive, recognise or articulate. Fourth, a geopolitical event challenges existing ways of making sense of the world, that is, an existing corpus of representations, theories, discourses, techniques and narratives through which the world is understood and practiced. Highlighting sensation, sense-making and the relations between them allows us to understand geopolitical events as being aesthetic as well as political. Geopolitical events are significant in that they may shake up and alter the distributions of sensibility (Rancière 2006), structures of feeling (Raymond Williams 1977) and imaginative geographies (Said 1978) that have coalesced out of previous events, while also giving rise to efforts to repair and reinstate them. Events have aesthetic, epistemological and ontological, as well as political and ethical, dimensions and implications.

This is already to go too far, however. Setting out a conceptualisation of 'the' event before working through 'an' event might do too much to frame our understanding in advance. It also risks divorcing the thinking of the event from the events and worlds amidst which it occurs. Here it is relevant to note that the recent critical thinking of the event in the humanities and social sciences has often been dominated by reference to a select group of male French philosophers working in Europe since the middle of the twentieth century, but ranges well beyond them.[2] In what follows, I link their thought to decolonial, feminist and anthropocenic accounts of the event, while also noting how these thinkers have been working, albeit from different locations, amidst and in relation to events, including the Second World War, the Holocaust, the detonation of nuclear weapons, the Cold War, and in relation to anticolonial struggles, as well as other kinds of revolutions, rebellions and liberation movements. The thinking of the event also flows through feminist analysis of the differentiated nature of invasion, occupation and war as events, and of the loaded nature of the categories, practices and institutions through which events are appropriated, represented and acted upon. More latterly, the thinking of the event passes through events like 9/11, the war on terror, Occupy, the Arab Spring, #BlackLivesMatter and the current decolonial moment, as well as the call, and struggle, to recognise the anthropocene as an event that has transformed, is transforming, and will transform, the world and ways of sensing and making sense of it at a fundamental level. Finally, as the book goes on to explore, a sense of events might also be altered by artworks that emerge from and reassemble them.

In the remainder of the chapter, I outline nine propositions for thinking geopolitics through the event. These propositions are neither comprehensive, nor independent of each other, nor necessarily definitive. Instead they offer orientations for thinking about geopolitical events, particularly as prompted by the artworks discussed in the book, to which the book as a whole and this chapter are themselves in large part a response. As has long been recognised across the humanities and the social sciences, and as the conceptualisation outlined above suggests, events pose myriad ontological, epistemological and analytical problems as well as ethical and political dilemmas. As Gilles Deleuze (2015, p.55) writes in *The Logic of Sense*, 'The event by itself is problematic and problematizing', and it is with Deleuze's thinking of the event that the chapter remains in closest conversation. As Deleuze's work would suggest, the thinking of any particular event must itself be 'evental'. This chapter and the book as a whole try to enact this principle.

Events Are Contingent

Events have often been understood as such to the extent that their 'eventuation' could not have been predicted or deduced theoretically, or extrapolated from their preceding or surrounding conditions. As Jacques Derrida (2007, p.451) has asserted, 'a predicted event is not an event'. Events introduce something new, unpredicted and unpredictable into the world. In this way of thinking, events do not 'emerge' 'in relation' to anything; they just occur. To the extent that it is possible to explain or even identify events, this can only take place retrospectively and never concurrently or in advance. An event is therefore distinct from the situation (Barry 2013) or context amidst which it erupts. As Andrew Robinson (2014, unpaginated; see also Dewsbury 2007; Shaw 2012) writes, summarising Alain Badiou's thinking, 'An Event interrupts the continuity of determinism. It allows something completely new to come into existence'.

If we privilege contingency, however, we might conclude that the 2003 Iraq war was by no means an event, given that it was widely predicted, and, apparently, readily explained as it began and continued to take place, as were its disastrous consequences. The 2003 Iraq war might seem to fail the test of radical contingency: immensely violent, disruptive and transformative, but a disaster foretold, many times over. I return to this point below, but my goal is not to adjudicate on whether or not some event or other is 'really' an event or not according to predetermined criteria; indeed, a paradoxical quality of events is that they can challenge our criteria for understanding what events are. Also, one can push the idea of the contingency of the event – or at least, of the geopolitical event – too far. Here Deleuze's thinking of the event is particularly useful. As Paul Patton (1997, unpaginated) suggests, drawing on Deleuze's work, there is often a certain patterning to the major or significant events with which we are often concerned, such as 'wars and treaties, colonizations and revolutions'. As he continues,

> These recurrent events are recognizable across variable circumstances and ways in which they occur in different societies at different times, but are not reducible to those circumstances, or to the bodies and material processes involved on each occasion. (1997, unpaginated)

Of course, not all events are recognisable in this way, or at least not initially, and one of the things that defines the most consequential geopolitical events like, for example, the Haitian revolution or the end of the Cold War, is precisely that they disrupt settled ways of sensing and making sense of the world; they seem to be singular. At the same time, I suggest, Deleuze's dual conception of the event as being both virtual (that is, partially expressing a certain deep patterning that is hard to grasp in everyday experience) and actual (expressed in unique, tangible, empirical form) offers a way of and thinking with and beyond contingency.

Events Are Shocking

Closely related to the idea of contingency is the almost axiomatic idea that the event is shocking and, related to this, surprising. The shocking and surprising nature of the event is expressed concisely by Jacques Derrida (2007, p.451), who stated, 'The event falls on me because I don't see it coming'. As Slavoj Žižek writes, 'at its purest and most minimal', the event is

> something shocking, out of joint, that appears to happen all of a sudden and interrupts the usual flow of things; something that emerges seemingly out of nowhere, without discernible causes, an appearance without solid being as its foundation. (2014, p.1)

Žižek thus ties the shock of the event to its contingency. As Derrida (2007, p.451) asserts, 'The event's eventfulness depends on [the] experience of the impossible. What comes to pass, as an event, can only come to pass if it's impossible.' People who encounter and survive events frequently relate their unreal or uncanny nature (for Deleuze, it is as if events enjoy an 'irreality', 2015, p.3), and their departure from normal experience is a central part of what seems to make them 'events'.

Here we might start to pick apart the embodied, affective shock of the event – its disruptive sensory character – from the more cognitively articulated experience of surprise. While we might be able to delineate these affects and effects in a phenomenological sense, however, we can also readily recognise how certain kinds of geopolitical event – indeed, one of the things that can make them 'geopolitical' – is that they involve a *differential* distribution of surprise and shock: I might not see the event coming, but maybe others did. A surprise attack is not a surprise to the people carrying it out (though they might be shocked by what it is actually like to go through it). This is one way in which an event can be understood as being multiple.

As historian William Sewell (1996) argues, an event is something that alters an established structure, and from this perspective it is the quality of 'structural' transformation that differentiates an 'event' from other kinds of occurrences. But while being forceful for some, certain kinds of events might not even be visible, audible or tangible within certain parameters of sensibility (Mirzoeff 2017). An individual or group might be aware of something taking place, but choose not to focus conscious attention on it, or push it to the margins of consciousness and attention. The event might also be welcomed and embraced, or denied and repressed, or experienced as a kind of harm. Events can also be encountered as holding promise, hope and potential as well as dread and horror (Anderson and Harrison 2010; Clark 2014).

The event, and extreme events in particular, have often been understood not just as shocking but as traumatic, as fundamentally undoing of self and world, and of the sensible and perceptual forms through which the two might be related (Edkins 2003). As Griselda Pollock (2013a, p.xvii; see also Jill Bennett 2005 p.3) summarises, 'the traumatic event exceeds cognitive recognition and the entry into conscious memory, while impressing its shapeless impact through aesthetic means: sound, hallucination, smell, colour and sensation'. The idea of a subject not fully present to itself or able to grasp the events in which it was caught up and produced has been formulated by a number of European philosophers since the eighteenth century (Eagleton 1990) and was advanced forcefully in psychoanalytic thinking in the wake of the industrialised slaughter of the First World War and thus, as Pollock (2013b, p.7) writes, in relation to the 'spatial, temporal and mechanical dimensions of technological modernity itself' (also Berman 1983; Doel 2017). In contrast to the eventally competent subject delineated in psychological research, the traumatised individual and collective subject of modernity is unable to grasp, process or move beyond the traumatising event, but is continually undone by its material, energetic and sensory force.

Maurice Blanchot expresses the traumatising and disorienting nature of the disaster, an overwhelming, ruinous event, in the following way:

> The disaster does not put me into question, but annuls the question, makes it disappear – as if along with the question, 'I' too disappeared in the disaster which never appears. The fact of disappearing is, precisely, not a fact, not an event; it does not happen, not only because there is no 'I' to undergo the experience, but because (and this is exactly what presupposition means), since the disaster always takes place after having taken place, there cannot possibly be any experience of it. (1995, p.35)

For Blanchot, the ontological structure of the disaster precludes subjecthood as well as knowledge and experience of it as 'an event' at all. The disaster, furthermore, has always-already happened: 'I', 'you', 'we' and 'they', as well as writing and knowledge, are compromised before we have really understood what is going on. In these respects, Blanchot's idea of the disaster both echoes and informs the idea of the pure or virtual event, or Event, whose deep patterning has already upset reality and causality, time and space, and the very possibility of subjectivity and knowledge.

The development of trauma theory was fundamentally shaped by the Holocaust, which Blanchot (1995, p.52) describes as 'the *absolute* event of history', and it has formed a central reference point for postcolonial and decolonial thought, while an interdisciplinary field of 'trauma studies' emerged in the post-Cold War period in relation to genocides in Yugoslavia and Rwanda. At the same time, problems have been identified in translating understandings of trauma derived from particular events and contexts into others, risking the depoliticisation of events and marginalising potentially beneficial practices (Leys 2010; Irene Visser 2011; Pollock 2013b).

The question of trauma returned to the forefront of Western political thought and practice amidst the event of 9/11, with the shocking, overwhelming nature that event rapidly appropriated in the effort to justify the infliction of violence via the war on terror. As the RETORT collective (Boal *et al.* 2005) argued, the war on terror and the 2003 invasion of Iraq in particular was launched in significant measure in order to provide a remedial counter-spectacle to the trauma of 9/11, understood as a symbolic blow to the authority of the sovereign state and US hegemony. As some critical theorists argued (Boal *et al.* 2005; Klein 2007), the sublime spectacle of sovereignty, expressed for example in the strategy of 'shock and awe' at work in the 2003 invasion of Iraq, has been employed since 9/11 as a political technique to reassert state power and facilitate the kind of economic restructuring that would not have been possible under 'normal' conditions. At the same time, critical reflections (e.g. Edkins 2003; Debrix 2008; Clark 2014) have suggested that the political instrumentalisation of shock events is fraught with ambiguity and uncertainty.

Events Are Entangled

Events are often understood as distinct, unique entities. In particle physics, the term event points to a specific, singular point in time-space, while Alain Badiou (2012, 2013) sees events as erupting at punctual 'event sites'. As Foucault (2000, p.381) writes, events must be understood in their uniqueness; genealogy 'must record the singularity of events outside any monotonous finality'. However, as he also writes, [t]he world ... *is a profusion of entangled events* (Foucault 2000, p.381, emphasis added). To think about events as being both singular and entangled, it is useful to draw on Deleuze's account, in which events are understood to be both individual moments of intensive transformation, and (in principle) infinitely extending, and multiply entangled, series (Colwell 1997; James Williams 2009).

Deleuze distinguishes between the realms of virtual potentiality (or Events) and empirical actuality (events or accidents), and as he suggests, the pure or ideal event, or Event, '*is not to be confused with its spatio-temporal realization in a state of affairs*' (2015, p.22, emphasis in the original). As Deleuze asserts, the virtual potential inherent in any situation is never fully actualised in an empirical state of affairs, and ostensibly distinct actual events remain connected at the virtual level. As he writes, '[ordinary] [e]vents are ideational singularities which communicate in one and the same Event' (Deleuze 2015, p.56); elsewhere he states that 'each event communicates with all others, and they all form one and the same Event' (Deleuze 2015, p.66). Actual events are singular, but the virtual Event is inter-linked with others, in multiple interlocking series (Colwell 1997) or deep patterns (Protevi 2009).

While Deleuze's account of the event is formidably complex, avowedly paradoxical and at times obscure, Paul Patton (1997, 2006) has suggested that European colonialism might be considered in this way as an Event, a recurring, virtual pattern that has been actualised in myriad unique, but interrelated, instances around the globe over an extended period of time. As Deleuze (2015, p.55, emphasis in original) writes, 'If the singularities are veritable events, they communicate in one and the same Event which endlessly redistributes them, while their transformations form a *history*'. We can make a link here with the reaction of journalist Philip Kennicott (2004; also cited by Gregory 2006a, p.229) to the pictures of Americans torturing Iraqis at Abu Ghraib that were released in 2004: 'In different forms, they could be pictures of the Dutch brutalizing the Indonesians; the French brutalizing the Algerians; the Belgians brutalizing the people of the Congo.' Such occurrences might thus be interpreted following Deleuze and Patton as specific actualisations of the violent virtualities of colonialism and of the anthropocene. One challenge of any event would then be, as theorists including Clare Colebrook (2010) and John Protevi (2009) suggest, to discern the virtual arrangements that specific events might be actualisations of; a second would be to think through the entanglements between apparently singular events at a virtual level. This is where, I will suggest, following Deleuze but also Jill Bennett (2012; further, O'Sullivan 2005; Colebrook 2010), that artworks can be particularly suggestive.

Events Are Immanent

Events can be conceptualised in different ways in relation to different conceptions of space and time. A common way of thinking about events in analytical philosophy (Casati and Varzi 1996) is in relation to absolute space-time: the boundaries of events can be established temporally with reference to their beginning, duration and conclusion and spatially in terms of their extents. Events might thus be tracked and mapped in terms of their emergence, development and conclusion through three spatial dimensions and one temporal dimension. But this way of thinking about events depends in turn on particular epistemological and technical assemblages that enable a quasi-objective, third-person perspective on events, that is, a certain transcendence with respect to the events in which a particular actor is interested, whether this be from the vantage point of an historian or a security manager monitoring a building. An alternative set of issues is opened up if we consider events as being immanent. An immanentist understanding of events does not see them has happening 'in' time or space, but as 'incarnating' (to use a Deleuzian term) their own forms of temporality and spatiality (Buchanan and Lambert 2005).

Deleuze explores the issue of event-time in several passages in *The Logic of Sense*, drawing on conceptions of time in Stoic philosophy. He distinguishes, first, *Chronos*, time understood as a succession of 'interlocking presents' (p.64) that is 'filled up with states of affairs and the movements of the objects that it measures' (p.66). *Chronos* is 'the always limited present, which measures the action of bodies as causes and the state of their mixtures in depth' (p.64). If *Chronos* is time as present, the 'actual' time of events, 'the time of bodies and states of affairs' (p.5), this can be distinguished from *Aion*, the virtual time of the Event. While *Chronos* is time understood as a succession of interlocking presents, *Aion* is time understood as never present, but rather always infinitely divided between a past just disappeared and a future not yet here. As Deleuze further relates, '*Sometimes*.. it will be said that only the past and future subsist, that they subdivide each present, ad infinitum, however small it may be, stretching it out over the empty line' (p.64). As he further explains:

> This Aion, being straight line and empty form, is the time of events-effects. Just as the present measures the temporal realization of the event – that is, its incarnation in the depth of acting bodies and its incorporation in a state of affairs – the event in turn, in its impassability and impenetrability, has no present. It rather retreats and advances in two directions at once, being the perpetual object of a double question: What is going to happen? What has just happened? The agonizing aspect of the pure event is that it is always and at the same time something which has just happened and something about to happen; never something which is happening. The pure event … is never an actuality. (Deleuze 2015, p.65)

While the (actual) event is incarnated, the (pure) Event does not 'exist', but inheres and subsists, escaping into past and future. We can link this with the commonly narrated experience of major or shocking events, in which time seems to work differently, being experienced not as a linear sequence, but sometimes as being slowed down, or poised on the cusp of becoming. The expected progression of things is interrupted, and perception is altered.

Traumatic events may enact a complete rupture in time understood as *Chronos:* as David Campbell (2001) wrote after the 9/11 attacks, there was a sense that 'time is broken'. As Jenny Edkins (2003) has written with regard to events of extreme violence:[3]

> In the linear time of the standard political processes, which is the time associated with the continuation of the nation-state, events that happen are part of a well-known and widely-accepted story. What happens fits into a pattern. We know almost in advance that such events have a place in the narrative. We know what they are. In trauma time, in contrast, we have a disruption of this linearity. Something happens that doesn't fit the story we already have, but demands that we invent a new account, one that will produce a place for what has happened and make it meaningful. Until this new story is produced we quite literally do not know what has happened: we cannot say what it was, it doesn't fit the script – we only know that 'something happened'. (2003, p.xiv)

The immanent event subsists, undoing subjectivity until some kind of transcendence can be achieved. This sense of interruption and rupture has spatial as well as temporal dimensions: in an Event, not only time, but also space, can be broken, twisted and ripped open, seeming to work in a different way. Familiar places are rendered unfamiliar, the landscape changes, arrangements are reconfigured, 'inside' and 'outside' are reversed or collapse into each other, and things disappear and appear in ways that are hard to grasp. The event is not a moment in time, 'but something that allows time to take off on another path' (Colebrook 2010, p.57). In the course of the Event, space, time and bodies are radically reshaped.

The question of the immanence of the event can also be traced in writing on colonialism and the anthropocene. As Patton (1997, 2006) suggests, colonialism is a world-breaking and world-making Event; as Frantz Fanon (2001, pp.29–30) wrote in relation to the effects of settler colonialism, '[t]he colonial world is a world cut in two... [t]he zone where the natives live is not complementary to the zone inhabited by the settlers. The two zones are opposed, but not in the service of a higher unity' (Mbembe 2001; Mignolo 2009; Vazquez and Mignolo 2013). Here I want to emphasise the ways in which colonialism has been understood in a manner akin to an immanent virtuality, a condition that is ongoing and which might be actualised at any moment. As black studies scholar Christina Sharpe (2016, p.9) writes, 'In the wake, the past that is not past reappears, always, to rupture the present'.[4] Here the figure of the wake can be interpreted as evoking the multiple ways colonial violence continues to 'subsist' as well as 'exist', as both deep patterning beyond immediate touch and tangible actuality. Gregory (2004a, p.iii) introduces the idea of the temporal connectedness of things and the disruptive, resurgent quality of the colonial present via William Faulkner's aphorism that 'The past is not dead. It is not even past.' As the Haitian anthropologist Michel-Rolph Trouillot (1997, cited in Sharpe 2016, p.55) argues, 'Pastness ... is a position. Thus, in no way can we identify the past as past.' In Deleuzian terms, the past subsists in the present as an immanent virtuality (Colwell 1997). As Sharpe (2016, p.16) writes, 'to be in the wake... is to recognize the ways that we are constituted through and by continued vulnerability to overwhelming force'.[5] For Sharpe, artistic practices are an important kind of 'wake work' (2016, p.59) operating counter to this violence.

The idea of colonialism as a deep or virtual patterning (Protevi 2009) that is actualised in empirical reality is also suggested by Patrick Wolfe's (2006, p.388) argument with reference to settler colonialism that 'invasion is a structure not an event'. This structure involves the establishment of a new form of territoriality via the dissolution of pre-existing forms, entailing what Wolfe calls a 'logic of elimination' in regard to the native population and the establishment of a new society, as the settlers 'come to stay' (Wolfe 2006, p.388). Wolfe (2006, p.388) clarifies his argument when he says that 'elimination is an organizing principal [sic] of settler-colonial society rather than a one-off (and superseded) occurrence'. The issue here, then, is not whether colonial invasion is a 'structure' or 'event', but the distinction between the immanent, virtual, patterning Event, and particular, empirical events.

The question of immanence is also raised in Blanchot's account of the disaster, a specific kind of event in which time, space and subjectivity are already ruined. Nigel Clark (2014) and Joshua Schuster (2014) have both drawn on Blanchot to think about the anthropocene, which, as Clark notes, calls for a fundamentally reshaped account of geopolitics. For Blanchot, the disaster has always-already happened; as Ray Scranton (2015, p.23) has written, the civilisation that has given rise to the anthropocene 'is already dead'. Such a condition poses formidable challenges for knowledge, the imagination and practice (Dixon 2015). The immanent event precedes experience and is enacted through, rather than by, people; it flows through them, affects them, carries them away and undoes them. As the next chapter goes on to explore, one of the things at stake in art is its ability to extract and reassemble something of the event (Jill Bennett 2012), thereby allowing it to be encountered in a different way.

Events Are Elusive

The immanence of the event poses ontological and epistemological problems. For Deleuze (2015, p.6), events 'are not things or facts... We cannot say that they exist, but rather that they subsist or inhere (having this minimum of being which is appropriate to that which is not a thing, a nonexisting entity)'. Furthermore, for Deleuze (2015, p.22), it is essential that 'the event is not confused with its spatio-termporal realization in a state of affairs'. We may recognise events via intense material-energetic alterations of bodies and states of affairs, but the event itself is an incorporeal transformation: 'The event is not what occurs (an accident), it is rather inside what occurs, the purely expressed' (Deleuze 2015, p.154).

The inherent elusiveness of events to conscious experience and knowledge is widely articulated across different accounts of the event. For Blanchot, the disaster cannot be recognised in advance or while it is happening; he writes that it only takes place after it has happened; likewise for Badiou (2012) the challenge is of grasping an event retrospectively and remaining faithful to it. The question of elusiveness can be connected with Jacques Derrida's attempt to think through the event via Martin Heidegger's reflections on the concept of *Ereignis*, a term that can be rendered as both 'event' and 'advent' (Blanchot 1995). Here it should be recognised that Derrida returned to the problem of the event following the 11 September 2001 terrorist attacks in the United States. Following these events, Derrida fulfilled a commitment to make an academic visit to New York, during which he engaged in a dialogue with Giovanna Borradori (2003) in which the problem of the event was a central preoccupation.[6] As Derrida described in relation to *Ereignis*,

> The undergoing of the event, that which in the undergoing or in the ordeal *at once opens itself up to and resists experience*, is, it seems to me, a certain *unappropriability* of what comes or happens. The event is what comes and, in coming, comes to surprise me, to surprise and suspend comprehension: the event is first of all *that which* I do not first of all comprehend. Better, the event is first of all *that* I do not comprehend. (Borradori *et al.* 2003, p.90, emphasis in original)

Events are not just shocking and surprising, then. For Derrida, an event is something that comes and is undergone, but which also at a quite fundamental level entails a certain limit in which it cannot be grasped; the event escapes appropriation, remaining in the realm of affect and never making it into conscious comprehension. As Blanchot (1995, p.112) relates, *Ereignis* is 'the event which cannot be understood in relation to "being"'. We can never grasp or recover the event itself; as Derrida suggests, borrowing from eighteenth century empiricism, it may be known only imperfectly by the impressions that it leaves. There is much more to explore here in terms of how events come to be appropriated via diverse techniques, and art in particular, but before we get to this point there are further features of events to consider.

Events Are Indeterminate

There is a further paradoxical quality to events as they are understood by Deleuze. While events can be said to have a certain singularity, the question of what any particular event itself is, of when and where it begins and where it ends, of what it involves and what it affects, is impossible to resolve with finality. If we follow the Deleuzian conception of the event, we must understand events as being indeterminate (Lundborg 2009; Kaiser 2012) and, in a related manner, non-localisable (James Williams 2009). Patton highlights this point in asking us to consider, 'at what point does a war begin? At what point does it end, given that its effects linger in the social and political life of the countries involved?' (also Lundborg 2009; Kaiser 2012).

These points are echoed in certain respects by geographical writers who recognise war as a complex, multidimensional, rather than discrete, occurrence. As Colin Flint has written,

> *war has a broad historical and geographical scope that extends far beyond the events of the conflict.* War should not be understood as a historic event bounded by the declaration of hostilities and the moment of victory or surrender. Nor should its geographic extent be restricted to where the battles and operations took place, an atlas of movement of lines of advance and retreat. The causes of war and the way it is fought emerge from decades, if not centuries, of political relations, and have implications that extend into postwar politics, perhaps for a number of generations. (2016, p.3, emphasis in the original)

To try to understand any particular instance of war, then, we must range beyond the ostensible boundaries of the event. If the contingency of the event means that it is not possible to infer or explain its occurrence from the contexts, relations and intersections (Loyd 2009a, 2009b) amidst which it emerges, critical understanding requires that we try to trace its dimensions, temporalities and spatialities as thoroughly and extensively as possible.

Events are also indeterminate in that occurrences that seemed to be past may be reanimated, and events that appear to have been confined can spread and spill over, or break out and proliferate. To some extent, this feature of events is a corollary of the idea of the immanence of the event as a virtual pattern not confined to any one location, but which may be actualised as different time-spaces – we don't know quite where, when or how an event will take place or how it might recur. But in Deleuze's thought the indeterminacy of the event also stems from the process of actualisation, which takes place along the edges and borders of the event; as he writes, 'everything happens at the border... the more the events traverse the entire, depthless extension, the more they cut and bruise' (2015, p.10). Events proliferate by touching and drawing things into their ambit, in the course of which those things are altered; events must be thought in terms of topological as well as topographic time-space (Allen 2011).

We need to take this idea a step further, however, especially in the context of political events. The non-localisable nature of the event is well explored work on the state of exception, which shows how events spread not only along their edges, but via the reversal and blurring of borders and boundaries, of inside and outside, that is, via the production of interiority and exteriority, but also liminality. The state of exception is an 'ambiguous zone' (Agamben 2005, p.2) between the political and the juridical, and between law and life, but also, in the figure of the detention centre, torture chamber or black site (Gregory 2006a; 2006b) between the inside and outside of territory. As Agamben (1998 p.19; also Secor 2007) writes, the sovereign decision on the state of exception – the originary juridico-political event – 'traces a threshold' between inside and outside, 'on the basis of which outside and inside, the normal situation and chaos, enter into those complex topological relations that make the validity of the juridical order possible'. As Secor (2007, p.40) elaborates, this condition 'permeates and overflows the spaces where the law operates in its various modes'. As Agamben (1998, p.19) emphasises, 'the state of exception itself is ... essentially unlocalizable.'

The non-localisable nature of the state of exception is evident in the tendency for states of emergency to appear in a seemingly arbitrary manner, to be extended in time, space, and remit, and to become normalised. It is also evident in terms of the ways in which the implementation of a state of emergency draws in ever-greater numbers of people and things, and in terms of how the creation of one exceptional zone necessarily gives rise to others. Furthermore, the effort to contain or localise a state of emergency, or to challenge and cancel it, draws in a growing constellation of actors into the event, contributing to its proliferation. Those opposed to the event or who simply wish to know about it, or commemorate it, become, in the act of opposing, knowing and commemorating, in some way part of it. The question is what such practices might make of the event.

Events Are Multiple

We are getting closer to what I will conceptualise as techniques of the event, and also to the question of counter-actualisation. Before reaching these points, it is useful to draw out further the multiplicity of the event and to explain how events are multiple in multiple ways. If we accept Deleuze's distinction between the virtual and the actual, we can make a distinction between the idea of the ideal, incorporeal, virtual event, or Event, and the corporeal, empirical, accidental event. The ideal event, Deleuze writes,

> is a singularity – or rather a set of singularities or of singular points… Singularities are turning points and points of inflection; bottlenecks, knots, foyers, and centers; points of fusion, condensation and boiling; points of tears and joy, sickness and health, hope and anxiety, 'sensitive' points. (2015, p.56)

This is not a distinction between two different kinds of events, then, but of two dimensions of every significant event, 'between the event, which is ideal by nature, and its spatio-temporal realization in a state of affairs' (Deleuze 2015, p.57). There is the Event, of which other events are 'its bits and pieces' (Deleuze 2015, p.58). In these terms, any event has a dual virtual-actual structure, the elements of which are linked by the process of actualisation, in which some of the potential inherent to a particular set of arrangements is activated, while other aspects are not, a process that in *Difference and Repetition* Deleuze (2004) calls differentiation. In the actualisation of an event, time is split between actual, successive, time, *Chronos*, and the ungraspable time of the Event, *Aion*, and space is ruptured, twisted, broken and remade. This helps us to move beyond a single, ostensible sense of an event as a definite and discrete occurrence 'in' space and time.

There are other dimensions to the multiplicity of events, however. Events can be thought of as having a fractal structure, in terms of how zooming in to examine any particular ostensible event invariably reveals further complexities. Developing Deleuze's thinking on the event, Paul Patton writes,

> Events are complex in that they are always composed of other events, however minimal or momentary. They are structured both internally and by their relations to other events. They often involve long periods when it appears that nothing is happening, and then suddenly everything changes and nothing is the same as before. In this manner, a political regime may collapse or an epoch draw to a close. (1997, unpaginated)

So events may have an at least dual temporal structure and entail complex spatial transformations, but may decompose into multiple other events and even non-events, as well as periods of more intense, rapid change, both when we are in the process of encountering them, or when we examine them from something of a remove.

There is one further aspect to the multiplicity of the event to draw out. People and places are positioned and transformed very differently by the events that flow through them, and while the idea of the event as immanent and elusive in an ontological sense would appear to preclude direct experience of events, it is self-evident that people are not only affected by events in very different ways, but often relate and interpret them in different ways at different moments, even when they have encountered what is ostensibly a common occurrence.[7] As Deleuze (2015, p.103) writes,

> If the battle is not an example of events among others, but rather the Event in its essence, it is no doubt because it is actualized in diverse manners at once, and because each participant may grasp it at a different level of actualization within its variable present.

We can say that people encounter events differently depending on how they relate to what Foucault (1998, p.93) called 'the complex strategical situation of power' and to what Said (1994) refers to as overlapping territories and intertwined histories. These situations and geographies are themselves complex, but they would include the sense of events happening both 'here' and to 'us', and 'there' and to 'them', or to 'allies', 'enemies' or 'the innocent', as framed by the domestic/foreign binaries of states and nationalisms but also in relation to Orientalist and other patternings of time, space and culture, that is, in relation to multiple distributions of sensibility and structures of feeling. As feminist work on Iraqi political history and the 2003 Iraq war (notably Al-Ali and Pratt 2009; Al-Ali 2011, also Al-Ali 2007) has demonstrated, and contrary to the claims of female liberation articulated by male politicians in the US and Britain as well as in Iraq, the nature and experience of events can vary in quite fundamental ways depending on gender and other axes of identification and subjectification (also Dolberg 2010). Rather than denying any possibility of subject-centred knowledge and experience, taking account of diverse accounts of an ostensible event is both an ethical imperative and a necessary element in thinking about the event in terms of its multiplicity and its virtuality beyond any particular set of experiences.

Events Are Appropriated Via Techniques

One perspective on events is that of the actor who is sufficiently empowered to command or orchestrate them, or who is tasked with doing so: the strategist, tactician or overseer of events. Such actors may somehow be inside the event, but their position necessitates a certain transcendence: they do not just go with the flow, but may have initiated the event and seek to monitor and shape it, and not be carried away by it. Here I am more concerned with the experience of being affected by and being carried along and away by events over which one has limited control, and which one struggles to apprehend, comprehend and appropriate. I am concerned mostly with the event as in some way from 'below', as well as 'inside', but also with the ways in which people, regardless of their relation to events, strive to produce and maintain some kind of knowledge of them, that is, with what can be called techniques of the event. Techniques of the event are ways of rendering events amenable to reflexive knowledge and practice in the face of their contingent, immanent and disruptive character. While Derrida (in Borradori *et al.*, 2003, p.90–91) emphasises the ultimate unknowability of events, he recognises the 'irreducible' and 'ineluctable' character of 'movements of appropriation'. These practices or techniques cannot and do not fully 'capture' or 'represent' events in an ontological sense, but are real and consequential, having actual effects on what people understand events to be and on how they act upon them, and they can be more or less accountable to the events with which they are concerned. I suggest that an effective account of an event can only emerge, not on the basis of an assumed or prior objectivity, but via an empirically, ethically and logically plausible and accountable working through of its multiplicity, that is, genealogically. This will not be an objective account, in that it is ontologically independent of the event. An effective account of an event is itself an event, which partially transforms the event that it is an account of, and of which it comes to form a part.

Techniques of the event are also often dependent upon and made up of several other techniques: a court case depends upon the taking of witness accounts, the gathering of records, forensic inquiry, the hearing of expert testimony; the cross examination and arguing out of things in public; deliberation and judgement. There are obviously many other techniques of the event, which have different forms and relationships to the events of which they render accounts, including things like biographies and memoirs, documentaries, memorials, inquests and inquiries. Such techniques may work with verbal, visual, sonic, numeric or other material elements, individually or in combination, and render their products in more or less technical, and more or less qualitative or quantitative form. Indeed, academic research can often be understood as being made up of and serving to create specialised, reflexive techniques of the event (e.g. Kaiser 2012).

Though effective accounts of events are possible, the obstacles to them are often formidable. Geopolitical events are invariably contested, involving and evoking attempts to falsify, deny and disavow their nature, as well as to place them beyond the limits of public knowledge. To put it differently, geopolitical events are often constituted at least in part through what can be called, adapting a term from Proctor and Schiebinger (2008), agnotopolitical techniques.

Such problems are particularly acute in terms of the epistemology of events of extreme violence. Dori Laub, a child survivor of the Holocaust and, later, a psychoanalyst who interviewed and treated other survivors and their children, wrote of how, 'during its historical occurrence, *the event produced no witnesses*'

(1992, p.75, emphasis in the original). As he argued, 'Not only, in effect, did the Nazis try to exterminate the physical witnesses of their crime; but the inherently incomprehensible *and* deceptive psychological structure of the event precluded its own witnessing, even by its very victims' (Laub 1992, p.80, emphasis in in the original). As he elaborated, 'As the event of the Jewish genocide unfolded... most actual or potential witnesses failed one-by-one to occupy their position as a witness, and at a certain point it seemed as if there was no one left to witness what was taking place' (1992, p.81). Even though an extensive body of detailed evidence has come to exist concerning how the genocide was implemented as well its political, legal and material dimensions, Agamben (2000, p.12) writes that 'The aporia of Auschwitz is, indeed, the very aporia of historical knowledge: a noncoincidence between facts and truth, between verification and comprehension'. Even with the emergence of an extensive body of evidence testifying to the event, Western societies have struggled to comprehend and acknowledge what exactly had happened in their midst (Edkins 2003).

As Brian Rappert (2012) has argued, the political management of the 2003 invasion and occupation of Iraq by the British government involved fostering a state of unknowing and unknowability concerning the extent of Iraqi casualties taking place during the event. The classification of documents and information, and the denial of knowledge relating to events, often rationalised in the name of a striking a balance between publicity and security (Thomas 2015), form important political techniques for the management of public opinion and ongoing events. When the existence of hitherto inaccessible or obscure documents relevant to the understanding of important events are identified, they are often released only under legal pressure and then only partially and in redacted form, or disclosed amid a flood of other less significant items, or at inconvenient times (Rappert 2012). While the Hutton, Butler and Chilcot inquiries into aspects of Britain's Iraq war gathered thousands upon thousands of documents, they were framed and conducted in ways that obviated some outcomes and rendered certain kinds of findings more likely than others. Art has particular significance in this context, in terms of its ability to model the event in ways that intersect with, open up, and bear upon, those of other techniques.

Events May Be Counter-actualised

What I am working towards is a sense of art as a particular technique of the event, which is able to counter-actualise events, that is, to do something with events that is distinct and different from other kinds of techniques. But first it is necessary to consider the process of counter-actualisation. Here it is relevant to note how Dori Laub interpreted the project of gathering survivor accounts of the Holocaust, which he described, in the following manner, as helping to create, after the fact, the missing Holocaust witness, in opening up the historical conceivability (the retrospective conditions of possibility), of the Holocaust witness. The testimony constitutes in this way a conceptual breakthrough, as well as a historical event in its own right ... what ultimately matters ... is ... the experience itself of *living through*, of giving testimony. (1992 p.85, emphasis in original)

The giving of testimony – and the specific project of gathering testimony concerning the Holocaust – is one particular way of enacting the event that transforms the event itself (also Patton 1997). There is an intriguing parallel here with the discussion in Deleuze and Guattari (1994) of two ways of going through the event, in which they take inspiration from the writer Charles Péguy, who is also cited several times by Deleuze in *Difference and Repetition* and *Logic of Sense*. As Deleuze and Guattari summarise, according to Péguy 'there are two ways of considering the event':

One consists in going over the course of the event, in recording its effectuation in history, its conditioning and its deterioration in history. But the other consists in reassembling the event, installing oneself in it as a becoming, becoming young again and aging in it, both at the same time, going through all its components or singularities. It may be that nothing changes or seems to change in history, but everything changes, and we change, in the event... (1994, p.111)

Deleuze and Guattari clearly privilege the latter way of considering the event, which as they suggest in relation to Péguy and other thinkers, enacts an 'aeternal', untimely or 'inactual' form of time, a virtual time of becoming, but the two are perhaps not as opposed or distinct as they might seem. Kaiser (2012), for example, suggests that academic work can engage in the reassembly of an event, a reassembly that depends upon and reworks prior recordings of the event's effectuation. The publication of the Report of the Iraq (or Chilcot) Inquiry in summer 2016, an attempt to record the effectuation of Britain's Iraq war in history, to some extent reanimated the event as a live becoming within British politics, for a time capturing the attention of participants and observers, and opening up the apparent possibility (albeit limited) of reshaping what appeared to be receding into the past.

Conclusions

In this chapter we have considered how events may be understood as being contingent, shocking, entangled, immanent, elusive, indeterminate, multiple, subject to appropriation, and open to counter-actualisation. What I will argue, drawing on Deleuze and Deleuze and Guattari, and building on the work of Jill Bennett (2012), is that, when they are made with regard to an event, artworks engage in a form of counter-actualisation that allows people to enter into and go through the event in a new way. To put this slightly differently, we can turn to Colwell's (1997) discussion of genealogy and counter-actualisation in terms of series and serialisation. As Colwell (1997, unpaginated) explains, in Deleuzian terms 'genealogy functions by decomposing the particular series along which events have been organized in order to create a different series for those events'; alternatively: 'genealogy counter-actualizes events, returns to the virtual structure of events, in order to reactualize them in another manner.' Artworks are genealogical in terms of the manner in which they can be said to reassemble or re-serialise events.

To conceptualise the way in which artworks may counter-actualise geopolitical events, I introduce the term eventual assemblages. As I shall explore in the next chapter and subsequently in the book, this term allows us to get at the ways in which the relation of artworks to events is itself multiple, inviting careful exploration. What also needs further discussion and conceptualisation beyond the literature to date are the conditions and means whereby artworks are able to do this. It is towards these tasks that the next chapter turns.

Notes

1 The singularity or otherwise of the world, that is, the question of whether people live in *the world*, *a world* or *worlds*, is part of what is at stake in a geopolitical event and accounts of it.
2 Principally, Maurice Blanchot, Gilles Deleuze, Alain Badiou, Jacques Derrida, Michel Foucault and Jacques Rancière.
3 François Debrix (2008) discusses this process with reference to Kant's account of the sublime.
4 I am grateful to Lioba Hirsch for drawing my attention to this book.
5 She continues: 'though not *only* known to ourselves and to each other *by* that force' (Sharpe 2016, p.16, emphasis in original). In Sharpe's writing 'we' refers to Black experience in the wake of the enslavement of African peoples. While the event of Black enslavement remains fundamental to the configuration of the world and its peoples today, colonialism took different forms in different regions and places and has been materialised specific ways in Britain's relationship with Iraq.
6 Jurgen Habermas was also part of this project; of the two, Derrida had more to say about the nature of events. I am grateful to Rory Rowan, who drew my attention to this book.
7 The classic cinematic exploration of this phenomenon is Akira Kurosawa's 1950 film *Rashomon*.

Chapter Three
Artworks as Evental Assemblages

In this chapter, I critique existing academic writing on art and geopolitics and develop a new approach that foregrounds the question of the event. The chapter works through four related sets of issues. First, it considers recent academic work that has discussed art in relation to geopolitics, geopower and aesthetics in order to draw out issues of ontology and epistemology via questions of critique, affect and posthumanism. It then considers how we can navigate the relationships that are posited in this literature and beyond, between terms such as affect, representation, cognition, embodiment and practice. It then builds on the work of Jacques Rancière to argue that art can be conceptualised genealogically in relation to geopolitics via the idea of the *dispositif*. This helps us to think about how the idea of art that has dominated in Britain, which has been disseminated around the world in the last 250 years, and which appears to offer a means of appropriating geopolitical events critically, is itself already implicated in colonialism and other forms of power. The chapter then connects with the Deleuzian idea of counter-actualisation to consider in more detail how we can think of artworks (and by extension exhibitions) as *evental assemblages*, in that they can be thought of as emerging from events, as themselves constituting events, and as engendering and participating in further events. As will become clear, the term artworks refers genealogically to certain kinds of things that are made, presented and encountered, which *happen* and *take place*, with regard to the *dispositif* of art, and which may take forms beyond that of the individual material object.

Geopolitics and the Event: Rethinking Britain's Iraq War Through Art, First Edition. Alan Ingram.
© 2019 Royal Geographical Society (with the Institute of British Geographers).
Published 2019 by John Wiley & Sons Ltd.

This allows us to approach geopolitics through art, but also to explore how art and geopolitics affect and relate to each other in the course of events. This way of thinking moves beyond existing approaches to art in political geography, which have tended to consider how art may enact what is often implicitly a one-time critical disruption of technologies of power. My approach draws inspiration from the work of Jill Bennett (2012), who has explored contemporary art engagements with the event, drawing in particular on Deleuze and Foucault. In *Practical Aesthetics*, Bennett (2012) considers diverse artistic projects and works in relation to diverse events, whereas the artworks I consider are all involved with what is ostensibly a common event. Tracking artworks and an event through each other adds to the understanding of both as multiple rather than singular things.

Art and the Human

Recent academic work on geopolitics, geopower and aesthetics has outlined a number of lines of thought on the nature of art and its relations to politics in the context of broader debates in the physical and life sciences, the social sciences and humanities. Among the most influential arguments to emerge from these debates have been that the linguistic and cultural turns in the social sciences and humanities effaced the material and affective nature of life and politics, and that they have forwarded an anthropo- and andro-centrism that is complicit in colonial and ecological violence. Though such a concise summary risks scanting what are sometimes complex and nuanced arguments and positions, the dominant tendencies in social theory in the wake of these critiques have correspondingly emphasised the material, the affective and the more-than-human. But while these critiques have opened or reopened important lines of inquiry, conceptual and analytical problems arise from the ontologisation of affective, materialist and more-than-human positions. These problems have already received considerable critical attention (e.g. Papoulias and Callard 2010; Pile 2010; Leys 2011; Wetherell 2012; Gratton 2014; Ivakhiv 2014; Cole 2015; Joronen and Häkli 2016); here I draw out their particular relevance for work on art and geopolitics. Trying to think about art and geopolitics in ontologised materialist, affective and posthuman terms, I suggest, risks becomes limiting and incoherent.

The first area in which issues of affect, materiality and the posthuman are raised concerns work on art in relation to the geopolitics of militarism, war and security. The prevailing conceptual rationale in this work, I suggest, follows a more general narrative about the critical nature, effect and possibilities of art (see also Ingram 2016). This narrative centres on the experience of artworks as affecting. This is shown by the way in which artworks relevant to geopolitics are described as, variously, powerful, compelling, shocking or revealing (e.g. Gregory 2010a; Brickell 2012). The person encountering the artwork is moved by it in an embodied and emotional way: they are affected by it, and this leads to, or at least heightens, critical insight into some aspects of geopolitics. This narrative structure allows us to think of art in evental terms, as an encounter between work, people and world in which something happens, and it is relevant to my argument that art is able to engage geopolitical events in part because artworks and our encounters with them are themselves events, in which we participate.

Implicitly evental reflection on the question of how art engages geopolitical phenomena has also been offered by Alison Williams (2014) and by Louise Amoore and Alexandra Hall (2010). Williams discusses the work of Fiona Banner, who has engaged extensively with the technologies of British air power and their affinities with bird life, as evinced in the naming of Royal Air Force planes after particular species. In one landmark work, *Harrier and Jaguar*, Banner installed two such planes, disarmed and with their markings and paint removed and their metal bodies polished to a reflective sheen, in the Duveen Galleries at Tate Britain. Williams draws on Judith Butler's account of discourse as both material and performative to argue that such works disrupt the operation of geopolitical power, supplying a specific conceptual rationale for how the work relates to geopolitics that centres on the moment of disruption.

Amoore and Hall, meanwhile, discuss a number of artworks that deal in some way with borders and security. They approach borders and security, as well as the visual, installation and participatory art works that engage them, through the conceptual lens of the theatrical and thus, again, implicitly as evental. In the case of Mark Wallinger's *State Britain*, installed at Tate Britain in 2007, they also reflect on how the space of the museum is interrupted by a work that consists of the recreation (again, within the Duveen Galleries) of an anti-war protest that had been cleared from Parliament Square. The idea of both borders/security and art as theatrical is evental in that both rely on an idea of embodied performance that happens through time and in space and which can be interrupted. In the encounter with artworks, borders and security are experienced in a different way, opening up the possibility of critical insight. Amoore and Hall also make the argument that the political implications of art do not stem from direct political commentary but rather from the artful staging of the contingency of power.[1]

In these studies, we have a sense that artistic enactments of geopolitics involve embodied encounters that are both affective (they involve material objects and technologies and work on and through the body) and effective (in that they elicit critical political effects, at least for the viewer, and potentially others). What is implied is an embodied human subject that is capable of being affected and of responding cognitively to this experience and therefore, also implicitly, an anthropic, rather than necessarily anthropocentric, idea of both art and geopolitics.

An anthropic position is questioned by work in geography that has considered whether aesthetics and art are in fact uniquely human categories and practices (Dixon *et al.* 2012). Of particular interest here are behaviours of birds and 'living rocks', entities that create beautiful and complex assemblages that might be said to embody a kind of aesthetics (understood here as a capacity for embodied sensing) and artistry (in that their works are expressive and rhythmical). Important is a sense of art derived from discussions of earth and territory in Deleuze and Deleuze and Guattari, via which both aesthetics and artistic expression are understood as manifestations of an immanent liveliness that is by no means confined to humans.

This approach to the artistic and the earthly also informs a rethinking of art, aesthetics and geopolitics through a particular conception of geopower.[2] This approach comes into geography via the work of Elizabeth Grosz, who sets out explicitly to question the 'ontology, that is the material and conceptual structures, of art' (2008, p.1). Grosz begins her lecture series *Chaos, Territory, Art* by asserting that this is a metaphysical project, and not the kind of work that would be 'confirmable by some kind of material evidence or empirical research' (2008, p.1). It is argued that art is 'the creation of forms through which ... materials come to generate and intensify sensation and thus directly impact living bodies, organs, nervous systems' (Grosz 2008, p.4). Furthermore, it is argued in Deleuzian terms that 'concepts are by-products or effects rather than the very material of art' (Grosz 2008: 4) and that '[a]rt is the art of affect *more than* representation, a system of dynamized and impacting forces *rather than* a system of unique images that function under the regime of signs' (Grosz 2008, p.3, emphases added).

Rather than analysing them together, then, Grosz sets affect against representation and opposes forces to images and signs, and states that the former categories are at least anterior, and possibly causal, with regard to the latter ones. This replicates the binary reasoning that has been questioned in critiques of affect theory (Leys 2011; Wetherell 2012), while the way that Grosz seeks to substantiate it in discussing art created in Australia by women of Aboriginal heritage conflates three quite different kinds of things: expressive behaviour across species; the ontologies and practices of human societies developed outside of and prior to colonialism; and the kinds of things that are made by people called artists and which get exhibited in galleries and museums (see also Grosz 2011). This confuses rather than clarifies the issues at stake.

Discussion in geography is more nuanced and less categorical, with Dixon *et al.* (2012) foregrounding the ways in which non-human entities (bowerbirds and thrombolites) can be said to exhibit artistry, while backgrounding the work of a human artist, who selects, frames, narrates and presents non-human objects as art, in a gallery setting, to human audiences, which discuss them. Here there are two different senses of the term 'art' at work, one of which (material creativity and expressivity) is more-than-human, the other (a reflexive conceptual, linguistic, social and institutional activity) being distinctively human, or as I suggest anthropic. While Dixon *et al.* set out to push and question, rather than dissolve, the limits we place on ideas of art and aesthetics, and while recognising that material creativity and expressivity and the 'anthropic' cannot be fully disentangled, these distinctions remain significant for my discussion.

A related but distinct set of arguments is advanced by Kathryn Yusoff (2015), who questions existing understandings of human subjectivity and art via the idea of the geological. While drawing in certain respects on Grosz, Yusoff questions vitalist accounts of existence that fail to recognise the role of inanimate and inexpressive materials – especially mineralogical ones – which, along with bacteria and fungi, it is argued, are anterior and interior to human subjectivity, and which both constitute it and pull it apart. Attending to what is argued to be a geological excess in subjectivity, Yusoff suggests, might enable humans to resituate themselves with regard to the anthropocene as a geological event. Yusoff argues that this is possible via an engagement with Palaeolithic art, specifically the famous Lascaux cave paintings of Southern France (which employ a variety of mineral-based pigments) and the less widely known *Gwion Gwion* rock art in Kimberley, Western Australia (which has been created with, and is reproduced by, 'living pigments' containing bacteria and fungi). These are important reference points for thinking about art because they date from the period within the last 50,000 years when *Homo sapiens* had begun its rapid journey to planetary dominance, but well before recorded history.

Yusoff argues that what we have come to regard as human subjectivity and identity are better thought of in Deleuzian terms as contingent events taking place at the confluence of inhuman forces and materials. This carries ethical and political import in relation to the prospect of human extinction: through recognition of 'geologic subjectivity' we might be able to accept non-normative or queer arrangements of things and develop 'different relations with the earth' (Yusoff 2015, p.389). The argument is not against the idea of the human or of human exceptionalism *per se*, then, but rather for more pluralistic and contingent accounts that might enable us to 'think beyond the image of man in the Anthropocene' (Yusoff 2015, p.403).

Yusoff takes from rock art an affective conception of aesthetics as an experimentation with inhuman and non-human forces of becoming, which thereby provides a passage to anterior, interior and non-local sources of subjectivity, sources that are 'cosmological, microbiological and geomorphological' (Yusoff 2015, p.399). This is human art, but the human subjectivity to which it testifies should be understood as more fleeting, more contingent, less human and more ecologically implicated than has conventionally been imagined. Furthermore, in providing access to inhuman and non-human virtualities, geomorphic aesthetics might allow alternative forms of in/human becoming. Yusoff thereby offers a more flexible, more open and more evental sense of humanity as an emergent becoming. A crucial question for the present discussion is how we might go from the Palaeolithic to the Second World War contexts within which the Lascaux cave paintings were rediscovered and brought into European art history, or to the colonial contexts within which Aboriginal art was, like rediscovered Assyrian art, disallowed from being considered on an equal footing with European forms. This is where the idea of art as a historically and geographically contingent *dispositif* becomes useful.

Communities of Sense, Structures of Feeling

We can also approach the question of the ontology and epistemology of art through Rancière's (2009b, p.31) discussion of the formation of a distribution of sensibility or community of sense via 'a certain cutting out of space and time that binds together practices, forms of visibility, and patterns of intelligibility'. A community of sense is, first, 'cut out', that is demarcated and differentiated from other entities via boundary making in space and time. A community of sense is also *sensed*; it involves ways of seeing and being seen, and the condition of being see-able; Rancière also mentions speaking and being heard; as part of this process of community-forming and to this we might add feeling, touching, smelling and so on (Panigia 2009). The community of sense is, further, a community of practices. Rancière does not offer a definition of practices, but the way he talks about them is consistent with the definition offered by Schatzki (2001, p.11; also Schatzki *et al.* 2001) as 'embodied, materially mediated arrays of human activity centrally organized around shared practical understanding'. This definition also points towards the ways in which a community of sense is defined by patterns of intelligibility, that is, of recognition and sense-making.

For Rancière, the formation of a community of sense involves both 'cutting out' and 'linkage' (or 'binding'), a process he also conceptualises as partitioning, sharing and distribution. The distribution of sensibility is therefore active, processual, evental; the formation of a community of sense is an event. But a community of sense can also be understood as *orienting of* events, as shaping how events are experienced and narrated, how they are made sense of; as emergent entities involved in the world in an ongoing way they may also be disrupted, displaced, marked and transformed by particularly intense events. *Geopolitical* events are typically more complicated than this, however, in that they involve multiple, intersecting, communities of sense, in different ways, such that the ostensible event that they might be said to experience comes into question. Rancière (2009b, p.31) also states that the term community of sense does not 'mean a collectivity shaped by some common feeling' but affect or feeling is not in fact absent from his thinking. Rather, I suggest, he is here directing attention to the processes and practices whereby such feelings may be elicited and orchestrated (Ben Anderson 2014). His thinking is therefore consistent with, and implies, I suggest, an emergent rather than essentialist or substantialist account of affect and feeling as embodied and materially mediated, and involving sense-making as well as sensation, a conceptualisation open, for example, to the idea of a nation as a particular kind of community that is felt as well as 'imagined' (Benedict Anderson 1983).

Isobel Armstrong's formulation of a feminist aesthetic is relevant here. As she argues, 'the components of aesthetic life are those that are already embedded in the processes and practices of consciousness – playing and dreaming, *thinking and feeling*... [t]hese processes – experiences that keep us alive – are common to everyone' (2000, p.2, emphasis added). Drawing on the work of Gillian Rose, Armstrong argues that the aesthetic is best understood as 'ceaseless mediation', a process that 'endows language-making and symbol-making, thought, and the life of affect, with creative and cognitive life' (2000, p.2). The 'constitutive moment of the aesthetic' for Armstrong is 'the structuring movement of thought *and* feeling' (2000, p.17, emphasis added). While Armstrong problematically asserts a singular collective subject ('us'), she provides a vocabulary for thinking about affect, cognition and so on in inter-related ways.

Raymond Williams' (1977) discussion of structures of feeling is also useful, offering a way of thinking about affect and sense-making that is social, materially mediated, specific in relation to time and space, and emergent. As Williams writes, with this concept,

> [w]e are talking about characteristic elements of impulse, restraint and tone; spe-
> cifically affective elements of consciousness and relationships: not feeling against
> thought, but thought as felt and feeling as thought: practical consciousness of a
> present kind, in a living and inter-relating continuity. We are then defining these
> elements as a 'structure': as a set, with specific internal relations, at once interlock-
> ing and in tension. Yet we are also defining a social experience which is still in
> process, often indeed not yet recognised as social but taken to be private, idiosyn-
> cratic and even isolating, but which in analysis (though rarely otherwise), has its
> emergent, connecting and dominant characteristics, indeed its specific hierarchies.
> (1977, p.132)

As in Rancière's discussion of sense, a structure of feeling is neither purely 'affective' nor 'cognitive': it refers to 'thought as felt' and 'feeling as thought'. As with Armstrong, we find in Williams a way of writing thinking/feeling both apart and together. Margaret Wetherell (2012, p.14), however, is critical of Williams' use of the term 'structures' to discuss what she calls 'complex coalescences'. Drawing on the novelist Iris Murdoch, she focuses rather on how 'affective machines' such as 'guilt, confession, penitence and redemption' can 'draw people like magnets' (Wetherell 2012, p.15), 'herding us along like brutes' (Murdoch 1967 cited in Wetherell 2012, p.15). It is, however, possible to make too much of the term 'structures'. It is significant that Williams places this word in scare quotes, and I would suggest that what his discussion offers overall is an under-standing of such entities as emergent, only metastable and as being interwoven with practices.

While a structure of feeling has by definition acquired a degree of organisation, this is for Williams emergent and not yet articulated in a fully public way. It is, further, constituted by 'internal relations' that are 'in tension' as well as 'inter-locking' and arranged in hierarchies: a structure of feeling may embody the 'shared practical understandings' posited by practice theory but these are not harmonious or without conflict. As Schatzki (1996, p.196) writes, '[p]articipants in a practice are clearly not equal within the webs of coexistence opened there. They are instead separated, hierarchized, and distributed.' Structures of feeling – meaning also the ensemble of practices and material arrangements through which they are constituted – may acquire a degree of continuity, may come to be expressed publicly and then dominate and exclude; they may also fade away, becoming residual, or be shattered by an extreme event. Like communities of sense, then, structures of feeling are evental: they *are* events; they are orientations *through which* events are experienced and enacted, and they may be reshaped more or less dramatically *by* events. Events, furthermore, may foster, activate or dissolve particular structures of feeling.

We are now in a position to consider artworks themselves. As in Rancière, Williams accords art a particular importance and function in relation to structures of feeling, in that art works may express social content of a 'present and affective kind, which cannot without loss be reduced to belief-systems, institutions, or explicit general relationships' (1977, p.132). As in Rancière, art is a material-affective-semantic complex that emerges from practices, and which interrupts and mediates them in complex and specific ways. As Williams further describes,

> The idea of a structure of feeling can be specifically related to the evidence of forms and conventions – semantic figures – which, in art and literature, are often among the very first indications that such a new structure is forming. (1977, p.132)

Artworks may thus be expressive of currents as yet 'at the very edge of semantic availability' (Williams 1977, p.133). I will later suggest that some artworks played this role during the Iraq war in Britain, but here I want to note further affinities in how Rancière and Williams construe the possible political role of art. For Rancière, what is most interesting is how artworks constitute communities of sense while participating in their formation more generally, thereby affecting the way in which the world is divided up and shared out. Although he sees several problems and limits in art's ability to act politically (Rancière 2010), its possible contribution lies in its ability to alter sensation and sense-making in ways that might be taken up or corroborated in politics proper. For Williams, art may be expressive of an emergent structure of feeling that may later turn out to be significant in how politics is thought and felt. He gives the example of how, while early Victorian ideology diagnosed poverty 'as social failure or deviation', semantic figures in the writing of Dickens and Emily Brontë 'specified exposure and isolation as a *general* condition' of which individual misfortunes were 'connecting instances' (Williams 1977, p.134, emphasis in the original). Art's relation to events may thus be prefigurative as much as expressive, or virtual as well as actual. Before advancing this line of thinking, however, it is necessary to develop further a genealogical approach to art, reconnecting with debates on art in relation to questions of aesthetics and the human.

The *Dispositif* of Art in Modernity/Coloniality

Before considering how artworks might counter-actualise the event, it is necessary to explore further what art itself is and how we have come to recognise it as such. This is particularly important because much of the discussion of art in relation to geopolitics either assumes an idea of art, or conceptualises it in only one of many possible ways, or in some cases conceptualises it in a way that risks incoherence. I begin with Rancière's implicitly genealogical take on art and open this out into a consideration of art and aesthetics in relation to modernity/coloniality.

Rancière argues that the idea of art that prevails in the world today acquired its meaning via an ensemble of rationalities, techniques and forms of visibility that emerged in Western Europe in the late eighteenth century. As he argues,

> [t]here is art insofar as the products of a number of techniques, such as painting, performing, dancing, playing music, and so on are grasped in a specific form of visibility that puts them in common and frames, out of their linkage, a specific sense of community. Humanity has known sculptors, dancers, or musicians for thousands of years. It has only known Art as such – in the singular and with a capital – for two centuries. (2009b, p.31)

While it is possible to conceptualise art in relation to capacities for affective intensity, material creativity, symbolic representation or embodied play, Rancière draws attention more specifically to a series of developments in a particular time and place that allow something referred to as 'art' to become the object of reflexive practices and a distinct domain and type of experience. As he argues elsewhere,

> 'art' is not the common concept that unifies the different arts. It is the *dispositif* that renders them visible. What the term 'art' designates in its singularity is the framing of a space of presentation by which the things of art are identified as such. (Rancière 2009a, p.23)

Furthermore,

> what links the practice of art to the question of the common is the constitution, at once material and symbolic, of a specific space-time, of a suspension with respect to the ordinary forms of sensory experience. (Rancière 2009a, p.23)

Art, then, is 'a specific sensorium that stands out as an exception from the normal regime of the sensible' (Rancière 2002, p.134, fn.1). It is in virtue of this exceptional status – its non-instrumentality and distance from ordinary life and politics – that art comes to have the powers widely ascribed to it within Western thought, practice and experience. Even more strikingly from a geographical perspective, Rancière states that art has come to be known as '*a certain partitioning of space* ... [a]rt is not made of paintings, poems or melodies. Above all, it is made of some spatial setting, such as the theater, the monument or the museum' (Rancière 2009b: 31, emphasis added). Art is therefore not reducible to specific objects (artworks) or practices (art making) that would constitute it; they may be expressions of art or as Rancière puts it, 'a system for the presentation of a form of art's visibility' (2009a, p.23), but to grasp what art is, we need to think of it as a complex ensemble of things that makes possible certain kinds of embodied and cognitive experience that are nameable as such.

In this way, Rancière's way of thinking about art recalls Michel Foucault's description of the *dispositif* as

> [a] thoroughly heterogeneous ensemble consisting of discourses, institutions, architectural forms, regulatory decisions, laws, administrative measures, scientific statements, philosophical, moral and philanthropic propositions – in short the said as much as the unsaid. Such are the elements of the apparatus. The apparatus itself is the system of relations that can be established between these elements. (1980, p.194)

What Foucault was trying to isolate here was, as he said, 'the nature of the connection that can exist between these heterogeneous elements' while recognising that 'there is a sort of interplay of shifts of position and modifications of function which can also vary very widely' (1980, p.194). As he also said, the term referred to 'a sort of – shall we say – formation which has as its major function at a given historical moment that of responding to an *urgent need*' (1980, p.195).

Here Foucault was thinking particularly of the disciplinary form of power that emerged in Europe in the eighteenth century and was expressed in prisons, clinics and schools and the forms of philosophical and political reason associated with them, but Rancière invites us to consider art and aesthetics in such a way also. This is highly suggestive, because what he calls the aesthetic regime for the identification of art came to be associated not just with philosophical aesthetics but with certain kinds of institutions and architectural forms – schools, museums and galleries *of art* – and with particular forms of subjectivity and practice – artist, spectator, curator – that also flourished in Europe from the eighteenth century onwards, and through which it was materialised and enacted. The idea of art as a *dispositif* therefore has some things in common with both Arthur Danto's (1964) argument that the ontology and epistemology of art is formed by its theoretical context and with the institutional theory of art advanced by George Dickie (1974), whereby the status of art is conferred upon artefacts by an elite group of experts, while going beyond both in that the idea of art as a *dispositif* implies links with governmentality more generally (also Tony Bennett 1988, 2015).

A further issue arising from Rancière's discussion of art in the singular is that it may appear overly deterministic of what art is and what it can do, reducing it to the effect of a particular institution (museum) or form (monument) and reducing its politics to their properties. However, approaching art genealogically counters this problem: a *dispositif* is not fixed but is materialised and enacted in varying ways, and its form and effects can vary and shift over time and space. It may bear the effects of intentions but has no single designer or strategist, its effects are emergent and provisional, and it may change significantly as a result of events. The *dispositif* is orienting rather than determining (Legg 2011). Indeed, the emergence of the *dispositif* of art can be considered as itself an event, through which we continue to live (in highly differentiated ways). Second, Foucault describes a *dispositif* not in terms of the sum of its parts but the 'system of relations' established between them, which might in this case correspond to something like the relations between artists, theorists, critics, the art market and art museums, which have emergent qualities, have evidently changed over time, and which continue to morph. However, the elements still matter too, and they may serve as entry points for understanding the *dispositif*, and each may, in the course of events, and more or less dramatically, reorient what the *dispositif* is and how it works. The key analytical points are that, while asymmetric, neither the workings of these elements nor the relations between them are determined, that the elements and relations can be appropriated (or assembled) in diverse ways, that appropriations may alter the workings of the *dispositif* itself, and that this may even lead it to become something else.

Rancière (2009a) locates what we can call the birth of aesthetics in relation to the emergence of modern European ideas and practices of freedom and equality in the late eighteenth century, around the time of the French revolution. As he argues, however, two competing forms of equality are at work in this sense of aesthetics. First, 'the system of constraints and hierarchies' (Rancière 2006, p.37) that previously shaped what counted as art breaks down, broadening the field beyond 'fine art' and beyond proper subjects such as portraiture and landscape painting, but so too does the distinction between art and life. This means that art is opened in theory to anyone and can be about anything. Indeed, for Rancière, the field of art is – in principle at least – infinitely open and incipiently democratic well before the twentieth century.

Rancière identifies a second form of equality in art where, in Kantian terms, works of art are 'free from the forms of sensory connection proper either to the objects of knowledge or the objects of desire'; artworks 'were merely "free appearance" responding to a free play, meaning a nonhierarchical relation between the intellectual and the sensory faculties' (Rancière 2009b, p.37). While for Kant, this free play ended up affirming the capacities of the subject for reason, for Friedrich Schiller, upon whom Rancière also draws, '[t]he "aesthetic state" defined a sphere of sensory equality where the supremacy of active understanding over passive sensibility was no longer valid' (Rancière 2012, p.37). The aesthetic regime apparently allows more open, more playful and less regulated interactions among the faculties and appears to serve as a model for political freedom and autonomy. This fragile, playful interrelation between the communities of sense enacted by artworks and by politics proper is what, for Rancière, gives art its political possibilities.

Terry Eagleton (1990) has traced the emergence of European aesthetics in similar terms, but while Rancière wants to assert the continuing political potential of the aesthetic after Schiller, Eagleton is much more suspicious as to the possibilities it offers for the assertion of equality. Indeed, Rancière's project to a large extent seeks to recuperate the aesthetic from the kind of critique that Eagleton makes of it. This critique is, nonetheless, worth considering, not least because it also deals with the problem of the relation between affect, thought and practice (see also Ingram 2016).

Eagleton takes as his starting point the question of why European philosophy since the Enlightenment has assigned such a 'curiously high priority to aesthetic questions' (1990, p.1). While different philosophers accorded aesthetics different significance in their thought, aesthetic questions, he notes, have formed a recurring and often compelling focus of reference and reflection: there is a 'strange tenacity' to 'aesthetic matters' (Eagleton 1990, p.2). This is attributable, on the one hand, to the manner in which the realm of art appears to present a counterpoint to

> progressively abstract, technical … thought, providing us with a welcome respite from the alienating rigours of other more specialized discourses, and offering, at the very heart of this great explosion and division of knowledges, a residually common world. (Eagleton 1990, p.2)

At the same time, however, aesthetics also itself becomes the object of abstract reflection and theorisation. In a self-contradictory way (because such reflection contaminates or destroys autonomous experience), the aesthetic comes to be thought of as 'a kind of paradigm for what a non-alienated mode of cognition might look like' (Eagleton 1990, p.2).

Eagleton also considers how aesthetics gets connected with political thought and practice. As he further argues, where the aesthetic is concerned, 'there is a certain indeterminacy of definition which allows it to figure in a varied span of preoccupations: freedom and legality, spontaneity and necessity, self-determination, autonomy, particularly and universality' (Eagleton 1990, p.3). Crucially,

> the category of the aesthetic assumes the importance it does in modern Europe because in speaking of art it speaks of these other matters too, which are at the heart of the middle class's struggle for political hegemony. The construction of the modern notion of the aesthetic artefact is thus inseparable from the construction of the dominant forms of modern class-society, and indeed from a whole new form of human subjectivity appropriate to that social order. It is on this account... that aesthetics plays so obtrusive a role in the intellectual heritage of the present. (Eagleton 1990, p.3)

While Eagleton tracks the way in which struggles over the aesthetic work through the debates of the 1980s surrounding postmodernism, it is possible to trace them further in the polemics surrounding the political implications of 1990s art movements like relational aesthetics (Rancière 2010), and more recently renewed interest in political aesthetics (e.g. McLagan and McKee 2012). The recent profusion of texts, as well as artistic and curatorial programmes, on aesthetics and politics testifies to the extent to which the aesthetic has returned to the forefront of theoretical reflection on political struggle. For Eagleton, aesthetics is a bourgeois mode of thought that offers a promise of life and experience in common that ultimately cannot be realised under capitalism. However, it is further necessary to note with Isobel Armstrong (2000) that, in his critiques of European philosophers' fascination with bodies, emotions and passions, Eagleton rehearses rather than critiques patriarchal ideologies of sex and gender. It must be recognised that ungoverned aesthetic experience not only raised the spectre of unruly proletarians but of liberated women, as well as other subjects who were understood as needing to be governed otherwise than through liberal freedom, and who could aspire to neither aesthetic or political equality.

If the aesthetic *dispositif* of art is implicated in the formation of modern European capitalism, patriarchy and the nation-state, it is no less implicated in colonialism. Though European dominance has been challenged by many aesthetic and artistic movements, the most influential art institutions and events continue to be concentrated within, and remain dominated by, European and settler-colonial societies. But the implication of art and aesthetics in coloniality is deeper than this. As Rancière (2009a, pp.8–9) touches upon (but does not explore), the metropolitan institutions through which the aesthetic *dispositif* of art was constituted established themselves and prospered at the expense of the people and places they disavowed and subordinated, both in terms of how they were often financed from the proceeds of colonialism and in terms of how they extracted artefacts from colonised lands, which were then made available for exhibition and study in the metropole.

As Rancière notes, the objects that came to populate modern European museums came from three main sources: the rediscovery of Greek antiquities; 'paintings and sculptures severed from religious, seignorial and monarchical grandeurs'; and 'the imperial and revolutionary pillaging of objects from conquered countries' (Rancière 2009a, p.8).[3] Modern European museums both appropriated cultural artefacts from other people, and displayed them in ways that validated and propagated ideas of European superiority. In the museum, objects were organised by and presented for either the scientific or the aesthetic gaze, depending on whether they were classified as artefacts (in the case of the products of non-European civilisations) or art (for objects deemed part of the European canon). While such distinctions were sometimes challenged by experts and visitors, the managers of museums and galleries (a specific kind of museum for the presentation of art) strived to uphold them. The aesthetic *dispositif* of art did not just reflect the colonial matrix of power within which it emerged; it played a role in constituting and perpetuating it. Distinctions between artwork and artefact were ways of partitioning people and the world and policing the forms of sensibility and sense-making through which they might be known. The emergence and bifurcation of the modern European museum along the art/artefact distinction are also geopolitical events in the sense argued here, in that that they enact a disruptive transformation of the world and of ways of sensing and making sense of it; museums exercise geopolitical power to the extent that they then help to consolidate and maintain the new arrangement.

Decolonial thought presents a forceful response to the implication of art and aesthetics in colonialism. As Mignolo argues, the central concern of decolonial thinking is not colonialism but coloniality, understood as a matrix of power centred on, and emanating historically from, Western Atlantic powers.[4] This matrix is further manifested in mainstream Western epistemology, aesthetics and ethics, which together form a 'prison house of language' (Mignolo in Gaztambide-Fernández 2014, p.205). A third, related, dimension of the colonial matrix of power is formed by processes of 'erasure, devaluation, and disavowing of certain human beings, ways of thinking, ways of living, and of doing in the world' (Mignolo in Gaztambide-Fernández 2014, 1 p.98). The colonial matrix of power wounds the colonised in that, as Frantz Fanon (2001) argued, it subjects a racialized other to a condition of inferiority and less than full humanity (also Sharpe 2016). For Mignolo, the European aesthetic *dispositif* of art descended from the eighteenth century is part of this matrix of power and helps to make it work. As Quijano asserts, noting the fascination of many modern European artists with the aesthetic expression of other peoples,

> What the Europeans did was to deprive Africans of legitimacy and recognition in the global cultural order dominated by European patterns. The former was confined to the category of the 'exotic'. That is, doubtless, what is manifested in the utilization of the products of African plastic expression as motive, starting point, source of inspiration for the art of Western or Europeanized African artists, but not as a mode of artistic expression of its own, of a rank equivalent to the European norm. And that exactly identities a colonial view. (2007, p.170)

The 'decolonial option' involves delinking from Western modernity/coloniality, including its ideas of aesthetics and art. However, this project must proceed from a recognition that Western ways of thinking and doing have been dispersed around the globe and, Mignolo argues, are inculcated in anyone who has a secondary or higher education. The decolonial option must seek to work around this condition, recognising entanglement and placing it in the context of power differentials. Mignolo also asserts a set of basic human capacities, including the social organisation of skills and creativity, which have been captured and continue to be channelled by a dominant form of thought and institutionalisation. Mignolo's term decolonial aestheSis (with a capitalised middle 'S') thus 'refers in general to any and every thinking and doing that is geared towards *undoing* a particular kind of aesthetics, of senses, that is the sensibility of the colonized subject' (Mignolo in

Gaztambide-Fernández 2013, p.198). Decolonial aistheSis is a form of cultural production that accomplishes delinking from colonial aesthetics. In the decolonial project, people may make use of the categories of art and artist and exploit institutions like the gallery, museum and art market, not to achieve success within them but as a working towards delinkage, and thus a world where European forms are seen in their proper way as local rather than global. This goal of delinkage is thus somewhat different from Rancière's idea of dissensus. Dissensus might prefigure delinkage, but not accomplish it.

Having emerged in Latin America in the 1990s, the decolonial project has rapidly been taken up in a range of artistic and curatorial agendas (e.g. Petrešen-Bachelet 2015) and is of direct relevance to the relationship, more than a century long, between Britain and Iraq, which, while it has not been defined by settlement and elimination, expresses entanglements and power differentials of a related kind. Indeed, struggles to assert different forms of Iraqi identity and cultural production within and beyond this relationship have many affinities with decolonial thought and practice, which I will explore in subsequent chapters. But while my project is concerned precisely with the entanglements and power differentials of the Britain-Iraq relationship, and while it can seek to recognise the force and implications of decolonial thought, it cannot be a decolonial project in the terms set out by Mignolo. This is the case, first, because decolonial thinking involves not just a theoretical perspective but a particular locus of enunciation: it is theory done from what is conceptualised as the border (see also Tuhiwai Smith 1999). This border epistemology emerges from the experience of coloniality, from being on the border of colonial power and it involves an embodied as well as intellectual knowing. As Mignolo argues, a non-colonised body simply cannot know what coloniality feels like, and its understanding is thereby limited. To the extent that a non-colonised body 'feels' as well as 'understands' coloniality, it is an experience of privilege.

My position is that, while academics who embody colonial privilege in the metropole cannot enact decolonial thought, a genealogical approach demands an engagement with artworks, arguments and analyses created from within the colonial borderlands, while recognising that there will always be limits to this process. In this light, the following statement, which comes in a chapter on artistic, cinematic and literary responses to the Iraq war, is decidedly problematic:

> Although there is doubtless much to learn about the lives of those who live in the war zones, my focus here is on the way contemporary artistic treatments – which challenge the limits of perception – can bring the lives of those on the U.S. domestic front into focus and thereby affect the ways in which wars will achieve future imaginative presences. (Shapiro 2012, p.142)

This statement, in an ostensibly genealogical analysis, enacts its own unfortunate partitioning of the world. First, it enacts the distinction between 'those who live in the war zones' and 'the U.S. domestic front' that the analysis ostensibly seeks to challenge, overlooking the presence of large and diverse Iraqi diaspora communities across Western countries. There is then, second, an assumption that it is ethnocentric works that will bring future wars home, obfuscating works that operate within as well as against the very entanglement or border condition that the analysis otherwise seeks to highlight. Third, the statement and the analysis of which it is part also overlooks the kinds of institutional and exhibitionary *dispositifs* through which Iraq has been brought into the West and made visible within it, and which shape the ways in which imaginative presences are constituted in the first place.

Recognising the force of decolonial thought, furthermore, does not mean subscribing to a fully binary understanding of the world, or of aesthetics and politics. As Linda Tuhiwai Smith (1999, p.27) has written, the categories of coloniser and colonised 'are not just a simple opposition but consist of several relations, some more clearly oppositional than others'. As I will explore, Eurocentric ideas of aesthetics and art were challenged by colonial archaeological findings in what the British called Mesopotamia, which revealed antiquities of great beauty and sophistication. Along with prehistoric human cave art, ancient Assyrian and Babylonian artefacts became reference points for European modernist art, which in turn influenced twentieth-century Iraqi modernism, which was itself also engaging directly with the legacies of previous Mesopotamian societies that colonial archaeology (a project that itself relied upon Iraqi knowledge and labour) made available. European and Iraqi aesthetics became thoroughly entangled within and against colonial matrices of power. In this book, I seek to offer a fuller account of how such relations of entanglement, asymmetry and resistance were expressed, appropriated and reframed through artistic enactments of the war, in part by engaging more fully with artistic, curatorial, academic and activist work by people of Iraqi heritage than has often been the case.

Counter-actualising the Event

The aesthetic *dispositif* of art is materialised and enacted in a multitude of ways and, Rancière argues, continues to exert an orienting and framing influence not just on what art is taken to be and how it becomes legible as such, but on its political possibilities. This, I suggest, often gets overlooked in discussions of particular artworks, projects or exhibitions: as a *dispositif*, 'art' emerges as a set of material objects and institutions and a constellation of practices, and via their metacognitive framing as belonging to art. As Rancière argues, the political possibilities for art are shaped – but, I argue, not determined – in relation with the *dispositif* through which art comes to exist as such. Tactics flowing from these possibilities include seeking to transcend or destroy art a separate domain of experience, as in the case of early twentieth-century Dada, or of moving beyond it to reach an 'outside', as in the more recent case of socially-engaged practices. At the other pole to this is the tactic of asserting art's autonomy from life and politics, exaggerating its strangeness and difference. The critical possibilities of art, Rancière (2006, 2010) suggests, lie in a skilful blending of these two politics of aesthetics, such that art does political work without advertising itself as such. However, he further suggests, it is not really possible to carry this off in a sustained way within the aesthetic *dispositif* of art, first because one or other of the politics of aesthetics tends to collapse, or dissolve the other, and second, because since the late eighteenth century, the idea that artworks can translate artistic intentions into sensuous experience and then into sustained, material political effect is no longer credible. The 'suitable' work of art remains a 'dream' (Rancière 2006, p.63), albeit one that continues to be pursued in many different ways. Art may offer intense imaginative and affective experiences and elicit new ideas and performances, but in more fragile, fleeting, provisional and ambiguous ways than what is usually taken to be politics proper.

It is admittedly problematic, having worked through some of the key ideas of decolonial thought, to return to European theorists in order to frame the ontology, epistemology and politics of art and the aesthetic, but, I would suggest, the arguments of Rancière and Deleuze conceptualise in a useful way much of what decolonial thought targets with regard to the politics of aesthetics, while offering other important insights. If, on the one hand, Deleuze offers a way of thinking about how events might be detached from their seeming self-evidence and enacted in a different kind of way, Rancière offers a note of caution concerning the extent to which radical or revolutionary political ideals can ultimately be realised through art as well as a qualification of the possibilities it might actually offer. But as well as reading their work in relation to decolonial, feminist and anthropocene thought, their ideas need to be channelled through some further critical qualifications as well.

For Deleuze (2004, 2015; also, James Williams 2009), the event is a forceful, material-energetic actualisation of virtual potential, a moment of intensive change whereby energy is released and bodies (both human and more-than-human) and states of affairs are transformed, systems change shape and new things become possible. The critical potential of events, he further suggests, lies in a contrary movement from the actual (or accidental, empirical event) back towards the virtual (the immanent potentiality from which the actual event materialised). This is the counter-actualisation of the event. In the previous chapter, I suggested that a phenomenological approach to the event could lead us to think about events in terms of first-, second- and third-person perspectives: the event as I encounter and understand it; as you do; and as others do. In *The Logic of Sense*, Deleuze (2015) goes a step further than this to identify a *fourth-person singular* position (also Colebrook 2010). This idea is especially suggestive in the present context, first, in terms of how it draws on the thought of the poet and writer Joe Bousquet, who fought and was wounded in the First World War, and, second, in terms of how it frames the event in relation to war, wounding and death. By considering this we can proceed to a conceptualisation of artworks as evental assemblages.

As Deleuze writes, the fourth-person singular is expressed in the form 'one' or 'it' (in the French, *on*):

> How different this *'one'* is from that which we encounter in everyday banality. It is the *one* of impersonal and pre-individual singularities, the *one* of the pure event wherein *it* dies in the same way that *it* rains. The splendor of the *one* is the splendor of the event itself or of the fourth person. (2015, p.156)

This further helps us to understand how Deleuze sees the event as something that flows through, enacts, and possibly undoes, people and things: 'The event ... signals and awaits us' (Deleuze 2015, p.154).

This might appear puzzling or unhelpful when we come to think about the Iraq war or geopolitics more generally, as it seems to suggest that ostensibly powerful or privileged people might be able to disclaim responsibility for events as being beyond their control. But we also need to recognise how Deleuze discusses responsibility for the event. As he suggests, the challenge is 'to become worthy of what happens to us' (Deleuze 2015, p.154). While this still suggests a certain passivity (an undergoing of the event rather than its doing, after John Dewey, 2005), he also posits a responsibility for the event and a necessity to enact it in a particular way. Drawing on Bousquet, he suggests that an event must be embraced and willed, lest one fall into a 'veritable *ressentiment*, resentment of the event' (2015, p.153). There is, therefore, a challenge to grasp and become responsible for one's own events: the event 'is what must be *understood, willed,* and *represented* in that which occurs' (Deleuze 2015, p.154, emphasis added). The event flows through people and things, transforming them, but it is imperative to appropriate something of it, to shape it or oppose it. To counter-actualise the event is to confront it and make something else of it.

There are still some obvious potential problems with this way of thinking about the event, to the extent that it implies and demands a masculine, individualistic and heroic orientation towards the event, of grasping, seizing and acting, or of overlooking the reality of actual historic and ongoing struggles. Counter-actualisation might also work through mourning, grieving or remembering, or bearing witness, or enduring and surviving, or even haunting. But Deleuze describes in the following way what he, drawing on Bousquet, sees as being at stake in 'willing' the event:

> If willing the event, is, primarily, to release its eternal truth, like the fire on which it is fed, this will would reach the point at which war is waged against war, the wound would be the living trace and scar of all wounds, and death turned on itself would be willed against all deaths. We are faced with a volitional intuition *and a transmutation*. (Deleuze 2015, pp.154–155, emphasis added)

In willing and confronting the event, Deleuze suggests, there is the possibility of its transmutation, or counter-actualisation. What is particularly relevant here is the way in which Deleuze sees acting (albeit in terms of binary genders) as a way of counter-actualising the event:

> The actor or actress represents, but what he or she represents is always still in the future and already in the past, whereas his or her representation is impassible and divided, unfolded without being ruptured, neither acting nor acted upon. It is in this sense that there is an actor's paradox: the actor maintains himself in the instant in order to act out something perpetually anticipated and delayed, hoped for and recalled. The role played is never that of a character; it is a theme (the complex theme or sense) constituted by the components of the event, that is, by the communicating singularities effectively liberated from the limits of individuals and persons. … The actor thus actualizes the event, but in a way which is entirely different from the actualization of the event in the depth of things. (2015, pp.154–155)

This provides a sense of how an artistic performance can enact an event otherwise by taking certain distinctive or guiding elements from it – its singularities – and composing them in a different way (Jill Bennett 2012). Viewed in this way, the work is less a representation than an experimental modelling of the event. As Deleuze further writes,

> the actor delimits the original, disengages from it an abstract line, and keeps from the event only its contour and its splendor, thereby becoming the actor of one's own events-a counter-actualization. (Deleuze 2015, p.155)

There are still some problems, though. It is unclear how an audience, playwright or director, or the theatre itself, might come into this situation. In this sense, I argue, Deleuze's idea of art is insufficiently developed to account for what he is positing: in modernity/coloniality artistic counter-actualisation has largely come to take place via – or, more critically, be captured by – a 'specific sensorium' (Rancière 2002, p.134, fn.1) that he does not theorise. Nevertheless, Jill Bennett (2012) has argued that Deleuze's idea of counter-actualisation provides a useful way of thinking about some forms of contemporary art. As she writes, 'By the logic of this process, art 'abstracts' from the actualised event a virtual structure that stages, in a particular kind of way, an affective encounter … which is, by definition, relational and open-ended' (Bennett 2012, p.43). As she further argues, 'an aesthetic reconfiguration of experience … does not simply restore subjective experience to history but generates new ways of being in the event' (Bennett 2012, p.43).[5] Artworks are otherworldly and thereby bear a genealogical relationship to the events that they counter-actualise.

Bennett (2012, p.36) has considered the ways in which specific artworks can provide a 'means of inhabiting and moving through events' or a way of 'being "in" rather than "about" an event'. As she writes in relation to *Documenta 11*, an art exhibition taking place in Kassel in Germany in 2002 which took as its subject the terrorist attacks of 11 September 2001, but linked them with a wide array of other events and situations, '[a]n exhibition itself is always in part a live event, unfolding in the here and now: an orchestration of present affects rather than a retrospective analysis. It connects as it exhibits' (Bennett 2012, p.37). In order to avoid the binaristic, opposed idea of sense and sense-making that I have argued is a problem in some work, I would argue that the relation between art and the event is better framed here in terms of 'as well as' and not 'rather than'; events are narrated *and* enacted, linguistic *and* affective.[6] Despite its oppositions, Bennett's work provides a useful template for thinking about both art and geopolitics as evental and of the former engaging the latter in evental ways. Following Deleuze and Guattari, she argues that 'art extracts and reconfigures features of an actual event' (Bennett 2012, p.39) and thereby poses something of its virtuality.

Bennett also discusses Doris Salcedo's sculptural installation *Tenebrae: Noviembre 7, 1985*, a work from 1999–2000 that deals with a terrorist attack on the High Courts in Bogotá, as being concerned with 'a mode of inhabitation – installing oneself within the event – which of course does not unfold in linear time, but encompasses past and future, interminable nothingness and a relationship to multiple other events' (2012, p.42). Rather than advancing a personalised or narrative account of the event, such works, Bennett argues, 'construe affectivity as the material substance of an event itself, and the means by which we enter into a given event, state of affairs or situation' (2012, p.42).

Bennett frames her thinking about events through the idea of practical aesthetics, which as she writes, 'emerge from a disposition, an openness to what happens, the measure of which is being "in" rather than "about" an event' (2012, p.36). While I would again phrase this in terms of 'as well as' rather than 'rather than', this sense of 'being in' and 'going through' the event is an important part of how I see art as being evental. As counter-actualisations, artworks and exhibitions can be seen as in some sense *part of* the event that they may appear to be 'about' and which they seek to appropriate. But as well as being *in* and *about* the event, artworks and exhibitions can also be thought of as emerging *out of* events, and as themselves *being* events.

A discussion of the act of painting by Jean-François Lyotard (1992, cited in Pollock 2013a, p.xxvii) can also help us to grasp this multiple relation of art to the event:

> A distinction should be made between the time it takes a painter to paint the picture (the time of production), the time required to look at it and understand the work (the time of consumption), the time to which the work refers (a moment, a scene, a situation, a sequence of events: the time of the diegetic referent, of the story told by the picture), the time it takes to reach the viewer once it has been created (the time of circulation) and finally, perhaps the time the painting *is*. This principle, childish as its ambitious may be, should allow us to isolate different 'sites of time'.

While Foucault (2000, p.381) sees the world as 'a profusion of entangled events', Lyotard suggests that we may 'isolate' and decompose the event of a work of art into evental elements. Artworks can be thought of as having several evental dimensions and properties, which will vary with the form as well as content of the works as well as in relation to its institutional, discursive and spatial settings, and which can be explored in different ways. I conceptualise this complex, multiple relationship of artworks to events by thinking of them as *evental assemblages*. Artworks are assemblages in that they are provisional coalescences (or territorialisations, Deleuze and Guattari 2004) of heterogeneous material, semiotic, embodied and energetic elements, each of which potentially exists in multiple relations to their own comprising elements, to each other, and to the surrounding world. They are evental in that they can be thought of as emerging from events, themselves constituting an event, and as having the capacity to engender further events, as they are completed or activated by viewers, spectators or participants. The enactment of two or more artworks together in an exhibition or public space intervention adds further evental inter-relations with other works, institutions and sites and artworks may be still further extended by practices of appropriation and reworking. I would further argue that while the virtual properties of artworks may be especially evident in particular art forms, such as sculptural installation (highlighted by Bennett), they are by no means unique to them. I would also argue that it is possible to extend the analysis to encompass the experience of geopolitical events and the creation (or assembly) of works by artists (as well as their selection and framing by curators) without personalising them entirely or interpreting them exclusively through a subjectivist or biographical lens. While Bennett highlights particular artworks and curatorial agendas as particularly embodying of Deleuzian ideas of the event, I am suggesting 'the evental' as a way of exploring and questioning an ostensible event across multiple artworks, while examining critically the *dispositif* of art that allows them to exist as such.

Conclusion

As the book goes on to explore, artists from different backgrounds experienced the 2003 Iraq war and responded to it in many different ways, as did institutions charged with commissioning, collecting and exhibiting art. Art and artists formed part of the massive protests against the coming war that took place in Britain and other countries around the world in February 2003, and while many artists of Iraqi heritage opposed the invasion outright, others experienced conflicting emotions. Some British soldiers sent to Iraq were also artists, while other artists accompanied or visited British forces. Still other artists were moved to respond to the war by specific events that took place within it. Many artworks intersected with the broader aesthetic politics and political struggles surrounding the war, and some resonated beyond their immediate art world and across into the public sphere.

The diverse and often divergent ways in which the 2003 Iraq war was experienced lead me to term it an *ostensible* event in order to place any singular account of it in question. This is potentially problematic if it leads to a sense that it is impossible to forge common accounts of the event at all. But I do not go quite so far as this. My argument is rather that it is only on the basis of a much fuller sense of the multiple nature of the event that we are properly able to understand it as *an* event. This has often been lacking, I suggest, in official, journalistic, academic and popular accounts of the war. As I will explore, a consideration of artworld enactments of the war in terms of individual works or longer term, multi-artwork projects and exhibitions broadens and diversifies our sense of the war as an event, especially if we take a critical sense of their institutional contexts as well as historical and contemporary resonances into account. It also broadens our understanding of how the war has been not just experienced, but appropriated and made available to be encountered otherwise.

Notes

1   We can draw out further evental layers to *State Britain*, to which I return in Chapter Four: this artwork was an event that dealt with an event (the clearing of a protest) that was itself about and against events (British policy and interventions in Iraq and Afghanistan).

2   These issues are also framed in Yusoff *et al.* (2012), Dixon (2015) and by Angela Last on geopower and geopoetics (2015, 2017). An alternative, Foucauldian, take on geopower is developed by Ó Tuathail (1996).

3   Major British museum collections were often founded on the basis of bequests or gifts from wealthy gentlemen (Sylvester 2009).

4   Thus downplaying non-Western colonialism, whether historic or contemporary.

5   Albeit in the rather different register of feminist psychoanalytic theory, Griselda Pollock also describes her approach to artworks in evental terms: 'I remain with … artworks in order to encounter certain movements or pressures within them that I identify as *traces of trauma*: events or experiences excessive to the capacity of the psyche to "digest" and the existing resources of representation to encompass' (2013a, p.xvii). There is thus the traumatic event that exceeds conscious appropriation and representation; the artwork that bears traces of it nonetheless; and the event of encountering the work and the traces it bears. An evental orientation is also evident in Pollock's (2013a, p.xxvii) questions of art: 'What are its occasions and its temporalities?' In Pollock's (2013a, p.xxviii) psychoanalytic methodology, time is understood as 'layered, archaeological, recursive'.

6   Bennett elsewhere (2005, p.11) states that she is interested in 'tracing the conjunction of affect and cognition', but her writing consistently opposes one to the other, emphasising affect 'rather than' cognition. While Deleuze and Guattari (1994, p.164) famously asserted that 'The work of art is a being of sensation and nothing else', I interpret artworks as assemblages of sensation and sense-making. Without the element of sense-making, a work of art cannot be distinguished satisfactorily from other things. See also Lotz (2009).

Chapter Four
Geopolitics at the Museum

In May 2006, an artwork entitled *Blessed Tigris*, a six-metre tall sculpture by the Iraqi artist Dia Al-Azzawi, one of the foremost figures in modern Arabic art, resident in London since 1976, appeared in the Great Court of the British Museum.[1] The sculpture was a brightly coloured rendition of a Babylonian ziggurat, embossed around its base with the calligraphic text of a 1962 poem by Mohammed Mahdi al-Jawahiri, who had struggled against British rule in Iraq and successive authoritarian governments, and who had written movingly of the pain of exile (Antoon 2003). As translated from the Arabic by Hussein Hadawi, the concluding lines of the poem read:

> O blessed Tigris, what inflames your heart
> Inflames me and what grieves you makes me grieve.
> O wanderer, play with a gentle touch;
> Caress the lute softly and sing again,
> That you may soothe a volcano seething with rage
> And pacify a heart burning with pain.
> (in Porter 2006, p.8)

The work had been commissioned for the exhibition *Word Into Art*, which examined the role of Arabic script in art from the Middle East, and its resonances with past and contemporary events were made clear in essays written to accompany it by Neil MacGregor, the then director of the Museum

Geopolitics and the Event: Rethinking Britain's Iraq War Through Art, First Edition. Alan Ingram.
© 2019 Royal Geographical Society (with the Institute of British Geographers).
Published 2019 by John Wiley & Sons Ltd.

(MacGregor 2006a, 2006b). The ziggurat towered over visitors and its bright colours stood out against the pale stone surroundings. With panels in vibrant blue, yellow, red and green as well as black and white, the sculpture combined the colour blocks of modernist painting with an ancient architectural form, which, along with the text, set sensation and sense-making in motion with and against each other. Employing techniques and forms developed by Al-Azzawi throughout his career, the sculpture also invoked the themes of exile and loss as well as the anger at political violence that run through his work. As with the *lamassu* that appeared in Trafalgar Square in 2018, *Blessed Tigris* was a temporary intervention, enduring there for the few months of the exhibition, an event entangled with several others: the ongoing occupation of Iraq, the exile of its creator and many other, Iraqis and the loss and destruction of the cultural heritage of the country. Like the *lamassu*, *Blessed Tigris* can be interpreted as reassembling Iraq's aesthetic and political history and present in material form, constituting its own reality in opposition to and amidst the violence of geopolitical events.

Also like the *lamassu*, the appearance of *Blessed Tigris* counter-actualised not only the ongoing event of the war but the space within which, but also by the grace of which, it appeared. In so doing, it also reframed the Museum, albeit temporarily and fleetingly, as a formative component in the modern European *dispositif* of art. As I explore in this chapter, in appearing at the centre of an institution historically constituted in significant measure on the basis of objects taken from what British explorers, cartographers and scholars had called Mesopotamia, *Blessed Tigris* intersected not only with the war but with the colonial genealogy of the museum and its reflexive transformation amidst post- and decolonial practices. It is therefore an evental assemblage of geopolitical as well as cultural significance, or, rather, it enacts the complex, evental, implication of culture, aesthetics and geopolitics with each other.

Blessed Tigris is revealing of the ways in which artists of Iraqi heritage responded to the 2003 war, and of the ways in which their works and those of others have emerged around, within and via institutions that are not only fundamental to the *dispositif* of art and the constitution of the public sphere, but which have historically devalued non-Western forms of creativity and aesthetics. For much of the last 200 years, the region that became Iraq has been known in British public culture via the matrix of coloniality. Mesopotamia gained prominence in the British imagination as the location of the ancient civilisations of Assyria, Babylon and Sumer and of scenes described in the Old Testament. These imaginations were forged to some extent through the accounts of explorers, geographers and painters, but primarily in terms of archaeological and museological discourses, institutions and practices. Although ancient Mesopotamian art exerted a formative influence on European modernism (Bahrani 2014), and while Iraqi modernism in turn drew on both on its way to becoming the most influential Arab art movement of the twentieth century (Bahrani and Shabout 2009), modern Iraqi art was until relatively recently largely absent from British art museums.

In this chapter I explore the ways in which artworks emerged around, within and via museums in the weeks, months and years surrounding and following the 2003 invasion and the ways in which works and museums mediated each other in relation to the ongoing event of the war. In so doing, I begin to demonstrate the diversity of artistic counter-actualisations of the war, and the differentiated ways in which artworks assembled in relation to Iraqi and British experiences have tended to emerge into the public sphere, a process to which museums have been central. In the first part of the chapter, and drawing on an extensive interdisciplinary literature, I develop an account of museums as geopolitical institutions, focusing on the mutual constitution of the British Museum and Iraq. I then track the emergence of art works before, during and after the invasion and in the subsequent months and years, showing how intial responses largely overlapped with anti-war protests and took place largely outside major formal institutions, before returning to the issue of how the British Museum has framed, and been reframed by, artworks associated with Iraqi experience. In the final part of the chapter, I consider high-profile artworks and exhibitions associated with the National Army Museum, Institute of Contemporary Arts, Tate Britain and the Imperial War Museum, each of which has also been important in mediating the appearance of the war in the public sphere, but which were also to some extent appropriated and reframed by the works whose emergence they facilitated.

Museums as Geopolitical Institutions

The modern European museum, a central component of the aesthetic *dispositif* of art, is a thoroughly geopolitical institution. As a series of critical interdisciplinary interventions (e.g. Tony Bennett 1998, 2015; Karp and Lavine 1991; Barringer and Flynn 1998; Preziosi and Farago 2004; Pollock and Zemans 2007; Sylvester 2009) have explored, the museum is invested with space and power in many intersecting ways.[2] Museums are predicated upon the exercise of power, first, in that they entail the assembling of a collection of objects, which are extracted from some places and concentrated in another. In a further exercise of power, those objects are then studied, categorised and displayed for a public gaze. These modes of categorisation and display flow from and serve to constitute a particular spatial and temporal organisation of the world and people and other things within it. Museums may over time be more or less open to different kinds of people and practices. Museums emerge, then, as products and expressions of nation building, of racial, class and gender politics, and of coloniality, each of which they may both normalise and reproduce, and which become sedimented in durable institutional and architectural infrastructures. Museums have in turn often become the target of demands for access and for restitution of property, while also being mobilised as agents of cultural diplomacy. Museums thus bear multiple relationships to geopolitical events, which they express, enact and elicit. In constituting certain kinds of public knowledge and space, museums may also become spaces for deliberation, critique and experimentation, with the question of whether museums can be decolonised emerging in recent years as a focal point for such practices. Museums are thus important spaces for exploring colonialism in terms of its virtuality, actualisation and counter-actualisation. If we are to understand how artworks might figure in these processes, we must examine how they emerge around, via and within them.

Running through the political history of the modern European museum is a pervasive, defining distinction between artefact and art that reflects the gendering and racialisation of power, knowledge and aesthetics, and which continually breaks and makes the world (e.g. Fraser 2005; Ruth Phillips 2007; Bahrani 2013). On the one hand, museums are places geared to the display of *artefacts,* understood as material evidence to be approached through the practices of science, and in relation to which knowledge is generated; on the other, museums may be geared to the display of art, understood as the objects of aesthetic experience and thus a realm of affect, sensation and contemplation. If we are to understand art as in some sense a product of a particular kind of *dispositif,* it is one that works through museums of art, but also through the distinction between the art museum as a space for aesthetic experience and museums of artefacts presented as evidence, as things to be known rather than experienced. 'Art' becomes a sensorium that is distinguished from everyday life but also from the scientific discourses that structure the museum of artefacts, within definite geopolitical coordinates.

The emergence of the modern European museum is bound up with the development of national identity and state authority, and the extension of colonialism beyond Europe into other world regions. As Preziosi and Farago (2004, p.2) argue, the modern museum must be understood not just as representational of the world and its peoples but as 'an institution for the construction, legitimization, and maintenance of cultural realities'. As they further argue:

> the museum is not only a cultural artifact made up of other cultural artifacts; museums serve as theater, encyclopedia, and laboratory for simulating (demonstrating) all manner of causal, historical, and (surreptitiously) teleological relationships. As such, museums are 'performances' – pedagogical and political in nature – whose practitioners are centrally invested in the activity of making the visible legible, thereby personifying objects as the representations of their makers, simultaneously objectifying the people who made them and, in a second order reality that is part of the same historical continuum, objectifying the people who view made objects in their recontextualized museum settings. (2004, pp.4–5)

For Christine Sylvester (2009) writing in IR, the modern European museum must be recognised as an active player in the constitution of both the national and the international via its enactment of class, gender and colonial ideologies and practices. As Craig Clunas, an art historian who formerly worked as a curator at the Victoria and Albert Museum in South Kensington, has written,

The National Museum acts as a key site of promotion of the existence and validity of the state formation. It does so with particular force in that the discursive practices at the heart of the museum lay claim to scientific objectivity, to a transcendental mimesis of what is "out there." It can thus act with particular force to validate the claims to sovereignty and independence by proving through displays of archaeology and ethnography the inevitability of the existence of the actually contingent conditions that give it its very existence. (1994, pp.319–320, emphasis added)

In this way, the modern museum asserted itself as 'a transcendental fixed point which observes all other "cultures"' (Clunas 1994, p.320).

The world-making capacity of museums entails violence. As Preziosi and Farago (2004, p.5) write, '[a]ny museological collection is, by definition, only made possible by dismembering another context and reassembling a new museological whole'. As the art historian Huey Copeland has stated, in a discussion of the National Museum of African American History and Culture and the National Museum of the American Indian in Washington DC, but which is relevant to the colonial genealogy of the museum more generally:

It's worth remembering that while there's a lot of time, money, labor, and attention invested in these particular buildings – these monuments, these physical, material manifestations – such formations are always haunted by theft and death. (Copeland 2017, p.261)

An object exhibited in a museum is thereby transformed, becoming an artefact in virtue of its extraction from its original context and accession into a collection and then exhibition, in which the circumstances and manner of its extraction are rarely accounted for. Exhibitionary practices involve disclosure and framing, but also erasure and disavowal, repression and haunting.

If we can trace the birth of the modern public museum to the opening up of the Louvre to a 'people' after the French revolution, we must also see the modern museum as being intimately involved in the ongoing formation of constellations of governmentality, which, while capitalising on an array of 'exhibitionary disciplines' construct particular forms of visibility and seeing and orchestrate acceptable forms of conduct and disposition (Tony Bennett 1988, p.87; Sylvester 2009). Museums not only reflect and help to shape a sense of a nation and its place in the world, a sense of 'here' and 'there', 'us' and 'them', but also the parameters of legitimate knowledge and behaviour.

While they are oriented in particular ways, however, neither museums, nor the objects within them, nor the aesthetic *dispositif* of art function deterministically, but are liable to shift, to be unsettled and to be contested. Nevertheless, British museums were shaped decisively by British class struggle and colonialism. A perceived need to respond to the French revolution informed early proposals to develop public museums, as new social groups challenged the established upper classes. The defeat of France in the Napoleonic wars led to many new objects coming into British possession, while opening up large territories in the eastern Mediterranean from which many more objects were taken. In this period, the British Museum, which had been founded on the basis of a bequest from Sir Hans Sloane in 1753, was substantially enlarged and restructured, and a National Gallery was created, moving into a grand new building in Trafalgar Square in 1838. The South Kensington Museum (formerly the Museum of Manufactures, and from 1899, the Victoria and Albert) was established on its current site in 1857 and was joined in 1881 by its neighbour the British Museum of Natural History. Another key museum institution, the Tate, was founded at Millbank in 1897 as the National Gallery of British Art.

As British influence extended across Mesopotamia during the nineteenth century, so a wealth of objects flowed towards Britain, above all into the collection of the British Museum. In 1807, Claudius James Rich was posted to Baghdad as the first representative there of the British India Company, a move that reflected anxiety about expanding French influence and a desire to secure communications with India, but which also forwarded the close relationship between colonialism and archaeology (Bohrer 1998). In 1825, the Museum received its first objects from the region, a group of around 50 Babylonian artefacts that had been collected by Rich. Further exploration was fuelled by Orientalist fantasies and a belief that the region contained the sites of Biblical events, including the Garden of Eden.[3] While held to be manifestly inferior in cultural terms, Mesopotamia, or at least its archaeological heritage, was not therefore understood as entirely other to Christian Europe. As Zainab Bahrani argues,

> far from being a peripheral field that was only recently incorporated into art history and aesthetic philosophy, the ancient Near East was an indispensable component of the development of a concept of a recognizable aesthetic in European discourse. It is the other that permitted the construction of the self. (2013, p.516)

In the 1840s, Britain and France expanded their archaeological missions in Mesopotamia, both initially on a covert basis. British work was led by Austen Henry Layard, who in 1849 was awarded the Founder's Medal of the Royal Geographical Society '[f]or important contributions to Asiatic Geography, interesting researches in Mesopotamia, and for his discovery of the remains at Nineveh'.[4]

Layard exercised a formative, but not definitive, influence on nineteenth-century British understandings of Mesopotamia. Layard had written on a personal level that Assyrian art, as expressed most imposingly in the *lamassu*, was 'greatly superior to that of any contemporary nation' (cited in Bohrer 1998, p.342). In official communications, however, he deferred to the expert opinion of Museum official Henry Creswicke Rawlinson, agreeing that 'the Nineveh marbles are not valuable as works of art', being inferior to the works of classical

Greece and Rome (cited in Bohrer 1998, p.342). Assyrian artefacts were instead placed in the same category as paintings and sculptures from Egypt and India. As Bohrer (1998:, p.342) recounts (see also Larsen 1996), aesthetic appreciation of the objects in keeping with their putative beauty or perfection posed a threat to 'established aesthetic doctrine'. A further threat to elite expertise was posed by the massive popular interest and appreciation aroused by the objects. As Bahrani (2001, p.161) notes, 'the general public... was awe-struck'. Popular aesthetic appreciation of objects from outside established frames of reference was therefore perceived as politically threatening. Exhibited in Britain, Assyrian objects threatened to upend judgements and categories that served to express and maintain the social order at home as well as abroad: elite status was secured in part upon the capacity for aesthetic experience, experience that found its proper reference in the appreciation of the Eurocentric art canon.

The extraction of objects from Mesopotamia continued until the First World War, by which time French and British archaeologists had been joined by Germans and Americans, the latter funded by private individuals including J.P. Morgan and John D. Rockefeller Jr. (Bernhardsson 2005). While Germany had dominated activity in the early twentieth century as a result of favourable relations with the Ottoman Empire (Al-Gailani Werr 2013), Britain gained control of the region during the First World War, and the Victoria and Albert and British Museums urged the government to use British control to 'bolster British national collections' (Bernhardsson 2005, p.72). Percy Cox, Britain's top official in Mesopotamia, demanded that a sizeable consignment of artefacts prepared by German archaeologists was sent instead to Britain, and this was subsequently shared out among British and international institutions. In 1918, the Keepers of the British Museum sent a letter to the military authorities in Mesopotamia asserting that 'the British Museum is the central archaeological museum of the empire and that science, fully as much as political considerations, demands that its contents should represent the archaeology of the British possessions' (cited in Bernhardsson 2005, p.74; see also Al-Gailani Werr 2014).

After the war, foreign archaeological research and the extraction of artefacts was formalised through the work of British archaeologist, diplomat and spy Gertrude Bell, who had helped to foment rebellion against the Ottoman Empire in Mesopotamia. In 1918 Bell had received the Royal Geographical Society's Founder's Medal, for 'important explorations and travels in Asia Minor, Syria, Arabia and on the Euphrates'; a year earlier she had written, in a letter from recently-occupied Baghdad:

My room is charming, warm and comfortable, with some delightful rugs which I've bought here on the floor, and all the new maps of Mesopotamia pinned up on the walls. Maps are my passion; I like to see the world with which I'm dealing, and everyone comes round to my room for geography. (Bell 1927, letter of 15 November 1917)

Having supported the installation of the Hashemite King Faisal I in Iraq, Bell was appointed by him to be the director of antiquities for the new and nominally independent state, and she assumed a central influence over archaeological activity in the country. Working to create and develop a national museum, Bell was at the same time able to make arrangements that favoured British access, in part by asserting that the demands of science were better met by relocating certain objects abroad to institutions including the British Museum (Bell 1927; Al-Gailani Werr 2013).[5] Things shifted following formal independence in 1932, as, under nationalist pressure, the Iraqi government moved to restrict exports and sought to recover objects from abroad, albeit mostly unsuccessfully (Bernhardsson 2005). As one Iraqi newspaper asked,

> May we throw a glance at our small museum and compare its contents with the objects unearthed in this country which have found their way into the museums which have been sending excavation missions into this country and find out whether our share has been a fair one or otherwise? (*Sawt al-'iraq*, 1933, cited in Bernhardsson 2005, p.1)

The wealth of Britain's collections, and the dependence and restricted scope of Iraq's, were products of the relationship between the two. And yet, as Bernhardsson further argues, Western archaeologists of the 1930s

> both underestimated the scientific and intellectual potential of the Iraqis and mis-judged the extensive impact that their own activities had had on Iraqi society, which had helped foster a respect and interest in Iraqi ancient history among segments of the population. (2005, p.180)

Successive Iraqi governments moved to subject archaeology to oversight and control and entered into negotiations with Western powers for the return of artefacts, and an appreciation of Iraq's ancient cultural heritage went on to form a central reference point for the development of modern Iraqi art after the Second World War. While in many respects exploited and travestied by the regime of Saddam Hussein, Iraq's cultural heritage continued to be recognised as being of fundamental world-historical as well as national significance.

The US and British governments received numerous expert warnings not just of the political risks of invading Iraq but of the risks war would pose to the country's cultural heritage (e.g. Bahrani 2003, but see also Curtis 2009). These warnings were disregarded. US forces failed to establish civil order in Baghdad and did not initially protect museums, galleries, libraries or archives. Such institutions were rapidly subject to vandalism, looting and destruction; thousands of irreplaceable objects were destroyed, damaged and stolen, and unsanctioned removal of artefacts from archaeological sites flourished. In an act of egregious disregard, US forces caused significant damage to the remains of the ancient city of Babylon by using the site as a vehicle depot (Curtis 2004), while stolen 'artefacts' and 'art' alike began to appear for sale outside the country. To many Iraqis, the 2003 war marked a new phase in the destruction and extraction of the country's heritage, comparable to the Mongol sacking of Baghdad in 1258.

The Appearance of Iraq

Early responses to the war in art can be traced through distinct communities of sense forming principally around anti-war protests and around Iraqi cultural organisations and projects.

On 15 February 2003, successive protests against plans to attack Iraq culminated in what is generally considered to be the biggest peacetime demonstration in British history, when between 1 and 2 million people marched through central London against the war as well as in cities across the country and around the world. While the scale of protests was not sustained, and while they did not succeed in preventing the invasion, the 15 February 2003 protest and the British government's apparent disregard for it subsequently became a central reference point in asserting opposition to the war, and crystallised a structure of feeling regarding the wrongness of the war that endures into the present.[6]

The protests can be said to have enacted a distinct aesthetic, which both expressed and forwarded opposition to the war on terror more generally; alongside the forceful appearance of hundreds of thousands of people in public space were thousands of mocking, satirical and humorous placards, banners and sculptures; anti-war graffiti proliferated. Many of these works not only channelled popular culture through the protest, but reworked images from particular moments in art and political history. Art images widely reproduced in the protests included a reworked version of the Peter Kennard Cold War-era photomontage *Defended to Death* (see also Chapter 7) and *Wrong War*, a stencil-based image created by street artist Banksy featuring a grim reaper figure with an acid house smiley logo (an icon of 1980s rave culture) for a face. These and other anti-war works were subsequently exhibited at *Pax Britannica* at Aquarium in Shoreditch in 2004, and as well as being framed as a key moment in activist-art history, went on to be integrated into the art market as commodities as well: the inclusion of Banksy's *Wrong War* saw the value of a portfolio of works released from the exhibition reach £6,500 at auction in September 2017.[7]

A particularly visible contribution to the February 2003 demonstrations and subsequently was made by the distinguished artist and designer David Gentleman, a longstanding Labour Party supporter who resigned his membership over the war (interview with author). Gentleman was already widely recognised, having created many well-known designs for Shell, the Royal Mail and Penguin Books, but had also published a book of graphic works critical of the dependent position adopted by the UK in its Cold War defence relationship with the US (Gentleman 1987). As it became increasingly clear that the Bush administration was intent on going to war in Iraq, Gentleman felt that anti-war protests were failing to achieve visual impact, and submitted a set of placard designs to the Stop the War group featuring the single word NO in black capitals against a white background, with red, blood-like spatters. These designs were adopted, reproduced and used widely on the 15 February 2003 March, lending a clear visual identity to many of the photos of that event and providing a template for many subsequent protest designs.

While academic, political and popular accounts of the war often emphasise the 15 February 2003 protest, the period of heightened political intensity surrounding the invasion also saw extensive activity by artists, curators and activists of Iraqi heritage, which came into the public sphere in a variety of ways. Largely overlooked in conventional accounts of the politics of the invasion, this field of aesthetically-oriented practice gave rise to a series of active, critical appropriations of the event, which, furthermore, emerged from and built upon prior and ongoing projects.

A second communuity of sense coalesced around around artists and curators of Iraqi heritage. A number of small galleries owned by people of Iraqi heritage and showing the work of Iraqi artists had existed in London since the 1970s. An official Iraqi Cultural Centre provided one gathering place for artists and artworks and the independent Kufa Gallery in Westbourne Grove another, where Iraqi and British artists had worked together to create work responding to the 1991 war. In 1995, the London-based artist, curator, gallerist and writer Maysaloun Faraj initiated a major research and curatorial project on art in Iraq and the diaspora, *Strokes of Genius: Contemporary Iraqi Art* (Faraj 2001), which was exhibited at the Brunei Gallery at the School of Oriental and African Studies in late 2000. In early 2003, *Our Life in Pieces*, an exhibition of everyday objects and stories chosen by people from Iraq was organised by the activist group Act Together – Women's Action for Iraq at an arts centre in Camden, coincided with the run up to the invasion. As described by curator Nadje Al-Ali (2011, p.101), the exhibition, which had been in preparation for more than a year, aimed 'to dispel the idea that Iraq can be equated with Saddam Hussain'. Reportedly visited by more than 2000 people, Al-Ali recounted that this event

> became a gathering point for many Iraqis who feared for their relatives' and friends' lives. We felt incredibly helpless, sad and angry, but we also knew that we were making a very small contribution by reminding people in Britain that a military intervention affected real people in Iraq as well as people living in the diaspora. (Al-Ali 2011, p.102)

Immediately following the invasion, Iraqi curators and artists rapidly pulled together exhibitions at Deluxe in Hoxton (*Before. After. Now. Visions of Iraq*) and at Aya Gallery in west London (*Expressions of Hope*) and organised the *Black Rain* project via the Ark Space in Ealing. As I shall explore further in the next chapter,

Iraqi art and cultural production was by no means absent before, during and after the invasion, but, while receiving a certain amount of media coverage, did not generally break into the wider public debates or fundamentally alter mainstream discourses surrounding the war, and tends to remain invisible in orthodox political accounts of the event. At this time, news media did report an increase in public interest in Iraqi art and culture via such shows, but also in terms of an increase in visitors to the Mesopotamian and Assyrian galleries at the British Museum (BBC News 2003a), highlighting the continuing significance of this institution in mediating public understanding of Iraq.

Two major cultural institutions initiated significant responses to the beginning of the war. With projects led by people of Iraqi heritage emerging largely via independent and smaller institutions in the period surrounding the invasion, the Imperial War Museum, via its Art Commissions Committee, initiated a process to identify a suitable artist to create an official work in response to the war, eventually choosing, as we shall see, Steve McQueen, while the British Museum moved to the forefront of international efforts to support the recovery, restoration and protection of cultural heritage in the country, efforts that involved working with military, police and intelligence agencies as well as cultural institutions and NGOs amidst the ongoing war and parlous security situation (Sylvester 2009; Hammilakis 2009; see also, Curtis 2009). As such efforts continued, the recently appointed director of the Museum, Neil MacGregor, asserted the continuing significance of its collections in a way that was unmistakably geopolitical. As he wrote in summer 2004,

> The new interim government in Iraq will have to consider how it defines Iraq's identity. And it will be surprising if it does not turn, as every other government in the Middle East has turned, to historical precedents to define the wished-for future. There is nowhere better to survey those precedents than the British Museum. (MacGregor 2004; see also Sylvester 2009)

MacGregor did not, however, explain how the Museum had come to attain such an essential position with regard to Iraqi national identity. Similar issues were raised by the way in which MacGregor presented *Word Into Art*, a major 2006 exhibition surveying the importance of the written and painted word in art from the Middle East, which reflected the spatialising practices of the modern European museum more generally. The Museum had begun acquiring contemporary artwork from across the region (taken to include Iran and North Africa and located mostly in the Department of Asia) in the mid-1980s as part of a broader strategy of expanding its contemporary international collections (Porter 2006). As Porter (2006, p.15) explained, work was collected particularly in view of the extent to which it '"spoke" of the region and showed continuity with "Islamic" art', that is, in terms of its representative (artefactual) as well as aesthetic (or artistic) significance.

MacGregor framed this exhibition in ways that further revealed the Museum's continuing implication with geopolitical imaginaries. As he argued, the Museum had been founded in 1753 'in order to stimulate citizens – and not just British citizens – to think about the new world that was opening up before them … and … as a vehicle of self-examination' (MacGregor 2006a, p.51). While MacGregor stressed contemporary political events in Britain, in particular the then-recent Jacobite rebellion, he did not set out how the world more generally might be understood to have been 'opening up' at this time. He set out the role he saw the Museum as playing in relation to present-day politics and global change in the following way:

> this is a show which does not just teach truths about a foreign culture. For the varied cultures of the Middle East – Arabic, Turkish, Israeli, Iranian – are now more at home in London than any other city outside the region. … We have heard much about 'Londonistan', caricatured as a resort for suspect aliens. We have not heard enough about how the world's most diverse city provides vital intellectual resources for the reimagining and reconstruction of the region – on its own terms. (2006b, pp.50–51)

For MacGregor, then, *Word Into Art* had a didactic geographical and anthropological purpose ('teaching truths about a foreign culture'), but carried wider significance in terms of London's privileged importance as a centre from which alternative geopolitical futures might be imagined for the Middle East. While apparently trying to emphasise the cosmopolitan orientation of the Museum, MacGregor (2006b, p.51) also revealed its enduring debt to coloniality in his quotation of a statement by Gowin Knight, its first director, that the Museum had been intended for 'as well natives as foreign', implying a three-fold distinction at its origin: British, foreign, native.

This is not to argue that everything about the Museum is determined by its colonial heritage or by the discourses through which its value and significance have been articulated by its leaders. Indeed, as demonstrated in the discussion of *Blessed Tigris* with which the chapter opened, to the extent that objects held and exhibited by the Museum are considered to be artworks more than artefacts, they are endowed with the capacity to resist such discourses and to question their location in the institution and the institution itself, and they can thereby be understood as holding out the possibility of a decolonised world, no matter how distant this might seem.

The framing power and effects of the Museum were also evident in another important exhibition of works by contemporary Iraqi, Kurdish and Syrian artists, *Iraq's Past Speaks to the Present*. This exhibition, which took place in late 2008 and early 2009, showed works lent by artists and drawn from the Museum's own collection, each engaging in some way with the heritages of Mesopotamia and each enacting more less direct resonances with contemporary political events. As the next chapter will explore, artists whose works appeared in *Iraq's Past…* have diverse political experiences and orientations, and the exhibition represented a further significant moment in the institutional recognition of contemporary (rather than historical) artwork relating to Iraq. While a discrete event, however, the exhibition was presented as a complement to a larger and more conventionally museological show, *Babylon: Myth and Reality*, which focused on the kinds of ancient Mesopotamian artefacts around which the Museum is more fundamentally organised, and which continue to constitute the main basis for its international standing. By contrast, *Iraq's Past...* involved just eight contemporary art works and took place in a single room also displaying Islamic objects framed as artefacts rather than art. At the same time, in presenting recently created works by contemporary artists that embraced and reworked aspects of Iraq's cultural heritage, often in relation to ongoing events, this exhibition can be interpreted as a critical and reflexive experiment with regard to the institution as well as events. Not only was Iraq's past speaking to the present; we might say that the works in this exhibition collectively posited the dream of a decolonised future, via an institution itself colonial in origin. This complex evental quality goes beyond framing in terms that are either didactic ('teaching truths') or representative ('speaking of a region').

These capacities are further evident in works from Michael Rakowitz's series *The Invisible Enemy Should Not Exist* that have been acquired by the Museum, and of which the *lamassu* that appeared in Trafalgar Square (discussed in the Introduction) is also an expression. Prompted by the looting of the National Museum in Baghdad, the project has attempted to reconstruct stolen artefacts that have been identified in databases held by the University of Chicago and by Interpol, using colourful packaging from Middle Eastern foods and Arabic language newspapers found on sale in the US. Sculptural works created by the project are typically exhibited alongside images and text concerning the original objects that they replicate. The project also includes hand-drawn sketches and texts by the author pertaining to the history and politics of cultural heritage in Iraq, and it takes its name from text relating to the Ishtar Gate in Babylon, which was excavated and removed to Berlin by colonial archaeologists in the early twentieth century, and then recreated *in situ* by the Iraqi government in 1950. As Rakowitz (2007) noted, while the occupation was ongoing, 'Today the reconstructed Ishtar Gate is the site most frequently photographed and posted on the Internet by US servicemen stationed in Iraq'. *The Invisible Enemy...* is thus a multi-layered and multidimensional project, which creates objects that stand in, imperfectly, for artefacts whose absence from their proper location is a consequence of the invasion, situating material culture in the context of colonial power, anti-colonial assertions of identity, and diasporic life.

The project has been exhibited in several museums around the world, including in the exhibition *Transmission Interrupted* at Modern Art Oxford in 2009.[8] In 2010, the British Museum acquired a series of objects created by Rakowitz and his assistants as part of the project, and in 2013 these were exhibited by the museum as part of a group exhibition in Room 34, the Islamic Gallery. These works included recreations of artefacts widely held to be definitive not just of Mesopotamian civilisations but the emergence of advanced human cultures (Bahrani 2014): cylinder seals, a jug, a plaque and sculpted representations of male and female figures and a bull. The presentation of this exhibition reflected the geographical frames for the Museum's collecting practices: while the name of Room 34 was 'Islamic World', the exhibition was described as presenting 'recent acquisitions of works by Arab artists' and seeking to 'highlight the breadth and range of art coming out of the Middle East in recent years, and the stories they tell' (British Museum 2017). While exhibiting works by cutting-edge contemporary artists, the ways in which these works have been collected, curated and exhibited therefore in certain respects continues to reflect longstanding museological discourses and practices that classify, divide and systematise the world according to broad categories, and museum records recognise elliptically that the 'particular relevance [of the works] to the British Museum lies in their resonance for the Museum's own collection and history as well as its strong historical and continued links with Iraq' (British Museum 2017). As well as bringing renewed attention to the significance of Mesopotamian material culture and linking to the Museum's ongoing efforts to protect Iraqi cultural heritage since 2003, these works also continue to prompt broader questions concerning the colonial genealogies of the modern European museum and what a more fully decolonised museum might look like.

Structures of Feeling, Distributions of Sensibility

I return to works created by artists of Iraqi heritage in the next chapter and throughout the book. For the remainder of this chapter, I focus on museum-mediated artworks that have attained considerable public prominence in British public life, and which can be said to have emerged in relation to the dominant structures of feeling and communities of sense through which the war was constituted as a public matter. The temporality as well as spatiality of the event is crucial here; while in some cases originating around the beginning of the war, it was only by 2006 and 2007 that works and projects began to emerge via large, publicly funded institutions, by which time structures of feeling concerning the event had taken on definite shape. A review of these shifts provides a sense of the context in which these works emerged and became part of broader debates and struggles surrounding the war. Relevant here are two further *dispositifs* through which the geopolitical event of war is both made a matter of public concern, and waged: body counts and opinion polls.

In the absence of counting by coalition forces, efforts to estimate excess or violent injuries and deaths due to the war among Iraqi people have produced widely varying results, but in any case massive numbers, from some 62,000 deaths (according to the Iraq Body Count project, based on reported deaths) to 650000 (according to a study published in *The Lancet* based on inference from surveys) (BBC News 2006a). The British government, meanwhile, sought to cast doubt on any such estimates and deflect from the issue (Rappert 2012).

More consequential for domestic politics were British military casualties, which also continued to accumulate, with more than 2000 troops injured by 2007 and more than a hundred killed. Several British citizens had been kidnapped and ransomed, and some executed. Scandals concerning torture and abuse committed by British troops had become public and their lack of protective equipment had become a major political issue. It had also become clear that ill-prepared, under-equipped and under-strength British forces had largely ceded control of Basra in the south of Iraq, for which they had assumed responsibility in 2003, to local forces backed by Iran. All of this combined with ongoing struggles around the original decision to go to war. In 2005 barrister and academic Philippe Sands (2005) published evidence indicating that the government's chief law officer had changed their view on the legality of invasion without UN Security Council authorisation, apparently, it subsequently transpired, in order to provide the legal cover demanded by leaders of the armed forces (see also Elden 2009). The Labour Party had gone on to win the General Election of 2005, but with a much-reduced majority and vote share and the leadership of Prime Minister Tony Blair had become increasingly beleaguered over Iraq in particular.

Public opposition to the war as measured by opinion polls also increased over time. Opinion polls should be interpreted subject to number of qualifications, but looking across several such exercises offers useful insights, especially concerning the difference made by the linguistic framing of the event. Over a period of months leading up to the invasion, one polling company asked a general question about whether 'the United States and Britain are/were right or wrong to take action against Iraq', producing support of 50 per cent through March 2003, rising immediately following the invasion to 66 per cent 'right' and 29 per cent 'wrong' in April 2003, shortly before Bush declared 'mission accomplished' (YouGov 2015).[9] Another company's question as to whether respondents 'would approve or disapprove of a military attack on Iraq to remove Saddam Hussein' immediately before the invasion, however, found only 38 per cent approval (Ipsos MORI 2003). Still more pointedly, a poll that asked whether respondents would support British troops joining an American-led action against Iraq in the event that 'UN inspectors do not find proof that Iraq is trying to hide weapons of mass destruction, and the UN security council does not vote in favour of military action' found only 26 per cent support and 63 per cent opposition (Ipsos MORI 2003). Had proof been found and action been approved by the Security Council, 74 per cent of respondents said they would have approved and only 16 per cent opposed. The same company found there to be net disapproval for Tony Blair's handling of the situation all the way from September 2002 through to the eve of war in March 2003.

In this context, Banksy's work *Wrong War*, reproductions of which had been prominent in the February 2003 protests, can be interpreted as an astute linguistic as well as visual intervention. A big drop in approval on the most general question ('right/wrong to take military action') from 58 to 50 per cent took place in May 2003, immediately following the death of the government weapons inspector

David Kelly, who had been pitched into the midst of the raging controversy over whether Blair and colleagues had 'sexed up' the infamous dossier on Iraq's supposed WMD programmes that was prepared in order to persuade Parliament to support the government's position. While approval on this measure recovered to 53 per cent in December 2003, it declined thereafter, with net disapproval consistently found from May 2004 onwards.[10]

Blair's credibility was undermined by his close association with George W. Bush, by his insistence on a clearly unfounded rationale for the war, and by his attempts to gain support for it via the infamous dossier. His rationale for the war was then dealt severe blows at several crucial moments, by the Kelly affair, and by the emergence of evidence that the decision to go to war and its implications for the risk of terrorism had been contradicted in crucial respects by the government's own intelligence and security agencies. In the run up to the invasion, Blair had argued strenuously that leaving Saddam Hussein in power was an over-riding risk that had to be addressed by military action, and subsequently defended his judgement in this regard. In September 2003 however the Intelligence and Security Committee (a statutory body established to oversee the work of the intelligence and security agencies) reported a secret Joint Intelligence Committee (JIC) assessment (*International Terrorism: War with Iraq*) of 10 February 2003 that had found that 'al-Qaida and associated groups continued to represent by far the greatest terrorist threat to Western interests, and that threat would be heightened by military action against Iraq' (Intelligence and Security Committee 2003, p.34). Neither the Hutton Inquiry into Kelly's death (which reported in January 2004) nor the Butler Review of the pre-war use of intelligence (July 2004) restored support for the war or Blair himself. While Blair denied any link between Iraq and the 7 July 2005 suicide bombings on the London transport network, *The Times* reported that the JIC had judged that the invasion had 'exacerbated' the threat of terrorist attacks in Britain, having 'reinforced the determination of terrorists who were already committed to attacking the West and motivated others who were not' (Leppard 2006). One of the bombers had recorded a video linking their actions with Western governments' violence against Muslims (BBC News 2005b); another had stated that attacks would continue 'until you pull your forces out of Afghanistan and Iraq' (BBC News 2006b). As we shall see, many artists strived to attach the war and the figure of Tony Blair to each other in as striking and indelible manner as possible.

This brief survey of opinion polls and surveys in relation to crucial events bears out a sense that major institutional responses emerged not just amidst negative and declining support for the war and growing opposition to it, but in the context of a profound – and, increasingly public – crisis within the state itself, as elements of the military, legal, diplomatic, intelligence and security agencies found themselves at odds with the position taken by Blair and his advisors. In this *milieu*, I suggest, museum-mediated artworks provided a crystallisation of, and focal point for, a range of emotions and views about the war, echoing, intensifying and shaping the structures of feeling to which it had given rise. These are especially interesting in view of the close relationships and overlaps that exist between the management of major cultural institutions and government and military circles. But struggles over the war were not just reflected in cultural institutions; these struggles were reframed and sharpened by the artworks and projects that emerged through them, and which in some instances continue to serve as salient reference points in relation to the event. The remaining sections consider the museum-related artworks or projects that drew greatest attention in this period, each of which was created by a male artist or curator already prominent in the British contemporary art world. While in many respects critical, in others these interventions also expressed existing patterns of privilege and prestige as well as a distribution of sensibility oriented around British experience.

Truth or Consequences, 2006

One expression of how major publicly funded institutions aligned with parts of the state sought to broach the issue of the war came in July 2007, with the move by the National Army Museum (NAM) to exhibit Gerald Laing's work *Truth or Consequences*, which, against the insistence of Blair, but in keeping with the intuitions and analyses of many other people and groups, made an explicit link between Iraq and the 7/7 bombings (Figure 4.1). The decision to exhibit this work was significant not just because of its intrinsic interest and timeliness, but in view of the close association between the Museum, the Ministry of Defence and the Army itself. The NAM was established under a Royal Charter in 1960 'to collect, preserve and exhibit objects and records relating to the Regular and Auxiliary forces of the British Army and of the Commonwealth, and to encourage research into their history and traditions' (National Army Museum 2016, p.3), and it receives around half its funding from the Ministry of Defence. Focusing on telling the story of the British Army, it is not a place where artworks that are so critical of British wars would usually be expected to appear, and its exhibition there itself constituted something of an event.

The exhibition of the work at the NAM gained further significance in light of Laing's own career, which illustrates the ways in which art practice, art history and geopolitics come to be intertwined in the course of events. Before becoming an artist Laing had been an officer in the Royal Northumberland Fusiliers and served in Northern Ireland. Having studied art at St Martin's College he went on to spend time in New York, becoming recognised as part of the pop art movement, creating paintings that rendered processed advertising imagery through newspaper photography-style dots and blocks of colour. Laing used pop techniques to respond to political events, including the Cuban Missile Crisis and the assassination of President Kennedy, but in the words of Morgan-Cox (2016, p.47), he came to be 'troubled by the paradox that the movement embraced the capitalist order it purported to critique'.

Figure 4.1   *Truth or Consequences,* Gerald Laing (2006). Images © The Estate of Gerald Laing. Reproduced with permission.

The catalyst for Laing's move into sculpture at the end of the 1960s was reportedly an encounter with the Royal Artillery Memorial located at Hyde Park Corner in London (Hodgkinson 2016). Constructed by sculptor Charles Sargeant Jagger and architect Henry Pearson after the First World War, the memorial was controversial in its time for its inclusion of the figure of a dead soldier, but has come to be regarded as one of the most powerful memorials to British war dead. It is further worth noting that the relief panels depicting scenes of intense fighting were influenced both by Jagger's own wartime experience and by Assyrian sculpture (Historic England 2017), forming another connection between British public culture and the legacy of colonial archaeology in Mesopotamia.

Laing was enraged by the 2003 invasion and its consequences, and returned to pop techniques to engage critically with the event, painting a series of works in 2004 and 2005 that culminated in *Truth or Consequences* in 2006.[11] These works collectively associate Blair and Bush with each other, with chaos and destruction in Iraq, with the torture at Abu Ghraib, and with previous British military interventions, as well as consumer culture more generally. Notable in these works are not just the employment of pop techniques, but references to the history of pop art, including a can of Campbell's Soup and a Colgate logo, as well as to Western art history more generally, with bodily forms borrowed from classical sculptural figures and from Grant Wood's iconic 1930 painting *American Gothic*. In this latter work, the instantly recognisable couple are replaced by Charles Graner and Lynndie England (low ranking personnel later convicted of abusing prisoners at Abu Ghraib), who (as in the notorious photo) are pictured raising their thumbs above a pile of Iraqi men whom they are torturing. These works mobilise a genealogy of their own art-historical form as well as the geopolitical event they engage and they are again exemplary of the central dynamic I identify regarding the complex and multiple ways in which art and geopolitics are implicated in each other, as well as the specific ways in which they have come together in relation to the Iraq war. The contentious content meant that Laing struggled to have these works exhibited in major public institutions and he consequently attached considerable significance to the adoption of *Truth or Consequences* by the NAM.[12]

Truth or Consequences is a painting on a concertina-like surface that reveals two different images when viewed from different angles, with one morphing to the other as one walks from side to side, thus realising a montage effect. Viewed from the left, the work places an image of George W. Bush next to a rendering of an image from footage of the bombing of Baghdad. When approached from the right, the viewer sees an image of Tony Blair next to a rendering of a widely-circulated image of the Number 30 bus bombed in Tavistock Square on 7 July 2005. It is also significant that a work by a former soldier was exhibited not just at the National Army Museum, but that this took place just a few days before the second anniversary of the July 2007 bombings, an event depicted in the work itself. The unveiling of the work was indeed criticized via parts of the media (Sawyer 2007) for the way that it appeared to link the bombing with the invasion.

Memorial to the Iraq War, 2007

Also contentious was the exhibition *Memorial to the Iraq War,* which ran at the Institute of Contemporary Arts (ICA) in May and June 2007. Taking place in the ICA galleries alongside the Mall, and thus close to the material and symbolic heart of the British state as well as many public memorials, this exhibition contravened normal memorial practices by inviting artists to make proposals while the event to be commemorated was still taking place. As it happened, Blair announced on 10 May 2007 that he would step down as Prime Minister, enabling Mark Sladen, the ICA's Director of Exhibitions, to link the thinking behind the exhibition to this development as well as to the event of the war more broadly:

> We have worked from a number of ideas when constructing this exhibition. One is that there always a need to consider how history might judge one's society; and for people living in Britain, the point at which Tony Blair leaves office is a pertinent moment to address the issue of our engagement with this war. Another idea is that by asking the 'wrong' question – by asking artists and visitors to attempt the impossible by stepping into the future and considering the notion of a memorial – we might gain new insights into our current situation; it is easy for those a long way behind the lines to become complacent about a long-running conflict. A third and important idea is that we are not seeking a definitive memorial to the Iraq War, but gathering different responses that will encourage debate about what can or should be memorialised from this terrible episode. (Sladen 2007)

This located the exhibition both spatially and with regard to an ongoing event, from the perspective of its imagined futures, an exercise through which critical insights, it was hoped, might be gained in the present. The artists who accepted invitations – including major established figures such as Jeremy Deller, Sam Durant, Lida Abdul and Jalal Toufic – created or proposed works of multiple forms and which bore more or less explicit relation to Iraq and to the event of the war.

The article written in the newspaper that was printed to accompany the exhibition in lieu of a catalogue made it clear that other artists had been invited to participate but had declined (Sladen 2007). And while this was not signalled in the exhibition materials, it is possible that some of the group did have some kind of link with Iraq itself. At the same time, it is nevertheless notable that none of the exhibiting artists was of Iraqi heritage or appeared to have had significant links with the country. To the extent that Iraqi art and artists appeared in major British museums by this point, this had taken place via the British Museum rather than institutions dedicated to art.

State Britain, 2007

A third important museum-mediated project, which illustrates the crossing over of practices and objects from activism into the art world while again raising the question of the (non-) appearance of Iraq in the public sphere, is Mark Wallinger's

State Britain, exhibited at Tate Britain in 2007, a work that has received some academic discussion (Bois *et al.* 2008; Amoore and Hall 2010; Murphy 2012; Ingram 2016).

In 2001 (and prior to 9/11 attacks), British peace campaigner Brian Haw had begun a continuous protest in Parliament Square, primarily against the sanctions regime then in force against Iraq. With the beginning of the war on terror, and then again after the invasion of Iraq, Haw had gathered more and more placards and other objects, including images of children and adults injured by military weapons, anti-war slogans, graphics and art works (including a work by Banksy showing British soldiers daubing a CND logo), as well as distressed clothing and dolls stained with blood-like liquid. Haw also lived continuously at the protest, sleeping for several years under a tarpaulin and then in a tent, until his death from lung cancer in 2011. Haw attracted support from a wide range of people for his determined opposition to US and especially British foreign policy, and the protest was widely taken to be a significant rebuke to politicians supportive of Britain's military interventions, receiving international as well as domestic media coverage. Several key politicians, including Blair, David Cameron and Boris Johnson, had expressed distaste at what they clearly felt to be an awkward and unpleasant spectacle that detracted from the grandeur of Parliament Square. Haw and his supporters upheld his right to protest in public space against a series of legal challenges.

In May 2006, police acting under a clause inserted in a new policing act concerned primarily with serious and organised crime cleared most of Haw's protest from Parliament Square, in order that it not exceed the now-mandated maximum surface area of three by two metres. While Haw returned to maintain a protest on a now much-reduced scale, the clearance prompted Wallinger to reproduce what had been cleared within the Duveen Galleries at Tate Britain, for which he had been commissioned to make a work. Having photographed the protest in detail, Wallinger spent several months overseeing the recreation of every item within it and had them installed in an identical layout in the galleries (Figure 4.2). As Wallinger stated, it 'seemed to be almost a … public service to try and make visible what has been rendered invisible' (Tate 2007). What had been rendered invisible (or at least, less visible), was protest against the 'war on terror' at least as much as the 'war on terror' itself, which was represented in a fragmentary and highly polemical manner in the protest. The work – like the protest – spoke as much to the question of the right of assembly and protest in a climate of heightened security policy as much as ongoing events in Afghanistan, Iraq and beyond (Amoore and Hall 2010).

This work, for which Wallinger won the 2007 Turner Prize, has been subject to contrasting interpretations (see also Bois *et al.* 2008; Ingram 2016). According to one line of argument (Murphy 2012), the recreation of the protest within the Tate simply confirms the extent to which contemporary art serves as a disciplinary mechanism, rather than politics by other means: while the protest had been

Figure 4.2   *State Britain*, Mark Wallinger, Duveen Galleries, 15 January–27 August 2007 © Tate, London 2018. Image courtesy of the artist.

dramatically curtailed in the name of 'security', what *State Britain* offered was art 'about' a protest rather than actual political confrontation. If politics is confined to the contemporary art museum, so the argument goes, the dominant order is indeed safe. A different line of argument (Amoore and Hall 2010) holds that this discordant and uncomfortable exhibition interrupted the experience of the grand Duveen Galleries, thereby prompting reflection on the part of visitors on the contingency of contemporary regimes of security. According to this argument, museum-based artworks remain important repositories and motivators of critical thought in the face of new forms of governmentality. In one argument, the quasi-autonomy of the work from politics diminishes its political value; in the other, it is this autonomy that enables its actual political function.

What I wish to highlight, first, is that such arguments can be seen to function via the parameters of what Rancière calls the aesthetic *dispositif* of art and second, that it is in eliciting such contrasting reactions (among others) that the aesthetic politics of the work may be said to lie. The inherent ambivalence of the work was therefore well captured by curator and critic Tom Morton (2007) when he wrote that while *State Britain* gave 'the often softcore stuff of participatory art a hard edge', it 'was at once a continuation of Haw's protest and something like a wake for it, an expression of freedom of speech and a wry, sad meditation on the art institution as a place in which dissent is contained and sanitized'.

What can be brought out more fully are the evental qualities of the work as an artistic recreation of a protest, and thus an event that is 'about' an event that was itself directed against an event (or constellation of events), to which the work also speaks. But such a framing would still be rather limited. It is also useful to extend our analysis to consider the broader relations between *State Britain* and the ongoing event of the war in Britain, and in particular the ways in which the work, which opened in January and closed in September of 2007, anticipated, framed, and endured beyond Blair's departure from office as Prime Minister. On 6 March 2007, Blair had given a speech in the Turbine Hall of Tate Modern in which he had emphasised the role of the arts in fostering human capital, social inclusion and national cohesion and in enhancing Britain's place in the world through cultural diplomacy (Blair 2007). Located a short distance from the core institutions of government, *State Britain* enacted a very different politics of art and can be seen as extending Haw's rebuke into the cultural realm, and beyond Blair's term. The status of the work, created by one of Britain's leading contemporary artists in one of its most prestigious arts spaces, also ensured that the work would become part of the way Blair's record in office would be interpreted and remembered.[13]

At the same time, while Haw's protest directly concerned the Iraq war, we can again note that it did not really represent Iraq or Iraqi people themselves and that in recreating the protest assemblage within the museum, neither did *State Britain*. This is not to suggest that they could or should have undertaken such work. Rather the point is that *State Britain* is illustrative of how major institutions and institutionally mediated works raised the issue of the war as a problem primarily of, for, and in relation to Britain, and thus pose the problem of the distribution of sensibility much more generally. While manifestly a critical intervention that provided a focal point for ongoing debates and struggles surrounding the war, Iraq itself remained somewhat in the margins. To the extent that it enters into the work, it does so negatively, as a constitutive absence.

Queen and Country, 2007–

Issues surrounding the marginal or non-appearance of Iraq and Iraqi people in the public sphere in Britain – and thus the nature of the event as a disruptive transformation of the world – are also raised by another important museum-mediated artwork, *Queen and Country* by Steve McQueen, which was also realised during 2007 and which also became part of the unfolding of the war in Britain. Tracking the inception, creation, form and reception of the work again sheds light on the multiple temporalities and spatialities of the war, which it came to play a part in shaping, and on how these were enacted via particular structures of feeling and communities of sense.

In early 2003, McQueen, already recognised as one of Britain's leading contemporary artists, was awarded a research commission by the Art Commissions Committee of the Imperial War Museum to create a response to the war, which, it was envisaged, would be exhibited at IWM London.[14] In that year, McQueen visited British troops in Basra, but was frustrated by restrictions on his movement and his inability to leave the British base. He subsequently struggled to formulate a work, but, mindful of the steadily growing list of British deaths, was later inspired while dealing with his taxes to propose the creation of a series of memorial stamps (Aspden 2007). Families of British service personnel killed in Iraq were contacted and invited to participate by contributing a chosen portrait photo of their loved one, each of which would appear, so it was envisaged, on a stamp along with the standard silhouette of the Queen's head in the top right corner. The title of the project thus referred to the way in which soldiers are held to serve both the monarch and the nation, which would be jointly embodied in the stamp. By 2010, 160 of the 179 British soldiers killed in Iraq had been added, in a remarkable show of support for the project among family members, who in the words of the artist became 'co-authors of the work' (McQueen in Aspden 2007, p.12).

As McQueen intended, the stamp form 'would be a better way to honour the dead than making some kind of three-dimensional object in London which no-one would come and see' (McQueen in Aspden 2007, p.12). The Royal Mail, however, declined to accept the proposal or put the stamps into production and McQueen also encountered resistance from the Ministry of Defence, recalling a meeting with officials at which he was asked, 'Why can't you do landscapes?' In response, the artist was at pains to stress that the work was not anti-war, but simply a way of honouring people who had died. Describing the process of working with relatives as 'difficult …very, very heavy', he also identified the antagonistic quality of the work: '[i]t is almost as if it was the first time they had been allowed to speak… These people have not been given a look in. They have not been treated correctly' (McQueen in Aspden 2007, p.12). In this context, *Queen and Country* articulated and amplified affective currents being encountered by the families of service personnel and wider publics who felt that the armed forces were being ill-treated and mismanaged, crystallising a particular structure of feeling in aesthetic form. In Rancière's terms, the project enacted an assertion of the equality of service members in opposition to the 'wrong' of their improper treatment by the state. Remembering them in this way, while it could not compensate for their deaths, would provide some kind of memorial, according them a place in national life; as McQueen envisaged, the medium of the stamp would 'enter the lifeblood of the country' (Art Exchange 2009). The project thus sought to mobilise an object that, while everyday or banal (Billig 1995) was nonetheless intimately connected to the formation of nationhood and, with the inclusion of the Queen's head on every stamp, state sovereignty.

The project was presented in a wooden stamp cabinet containing 120 double-sided vertical drawers, each side of which displays a sheet of stamps dedicated to one service member, accompanied by a handout listing their names and dates of death (Figure 4.3). Resting on a spare metal frame, the cabinet replicates the manner in which McQueen had found stamps to be stored in the British Museum

Figure 4.3   *Queen and Country*, Steve McQueen (2006) © IWM (Art.IWM ART 17290). Reproduced with permission.

(in Aspden 2007), while its form and mode of exhibition, in an otherwise empty room, lent it a distinct resemblance to a coffin on a stand. The act of carefully sliding out each drawer individually accords each person, whose face is displayed multiple times, a distinct period of attention; to view the whole work requires an extended period of time that fosters further reflection, drawing the viewer into the event of each individual fatality, each of which in turn forms a distinct element of the larger event of Britain's participation in the war.

The work was acquired by the IWM with support from the Art Fund but was first presented at the Manchester International Festival in June 2007, again corresponding closely with the departure of Blair from office, and was subsequently exhibited at a series of museums in England and in Northern Ireland as well at the Barbican arts centre in London, in conjunction with a performance of the play *Black Watch* (Burke 2007), which also dealt with British military service in Iraq. *Queen and Country* also became the focal point for a petition to have the stamps accepted and produced by the Royal Mail and the campaign received near-universal support from newspapers across the political spectrum. In 2010, the project was exhibited at the National Portrait Gallery and published as a book in collaboration with the British Council (McQueen 2010). McQueen has stated that he regards the project as incomplete until the stamps are issued.

Queen and Country constituted itself as a particular kind of evental assemblage, a participatory art project that achieved widespread public engagement and impact, sharpening discussions about the treatment and commemoration of British soldiers at a time when these issues were also at the forefront of political disputes and when other vernacular commemorative practices, most notably the spontaneous gatherings in the town of Wootton Bassett to mark the transport of the bodies of British soldiers following their return from Iraq and Afghanistan, were emerging (Jenkings *et al.* 2012). *Queen and Country* carried these issues via cultural institutions into the public domain and, enacting a temporary public, providing a platform for relatives to demand recognition from the state. The support given to the project by major public bodies including the Art Fund and the British Council further reflected the way in which the contested politics of the war were being refracted back through official institutions. The decision to go to war had created a particular structure of feeling characterised by profound unease and antagonism towards Blair's circle within parts of the state as well as across British society, which *Queen and Country* both elicited and intensified, without making explicit reference to the war as a political matter. If the work can be said to have achieved political effects, it did so precisely through an insistence on its autonomy from politics proper, that is, by exploiting the possibilities afforded by the *dispositif* of art.

Commissioned by the Imperial War Museum and exhibited there and in other major public institutions, *Queen and Country* became one of the most recognised and most widely-discussed art works to have emerged from the war, connecting experiences of the war and feelings towards it, in a way that blended form, affect and practice, enrolling people not just in viewing the work, but acting and feeling with and through it, in so doing helping to give form and intensity to structures of feeling. To understand how it has become a definitive artwork, we have to account for how it emerged via the mediation of major museums, including the IWM and the National Portrait Gallery, but also via other significant institutions, including those (the Royal Mail and Ministry of Defence) in relation to which it exists antagonistically. Like *State Britain*, *Queen and Country*, enacts neither social cohesion nor healing, but division and loss resulting from the war.

As with other high-profile works, Iraq itself does not really appear in *Queen and Country*, but nor is it entirely absent. The sheet containing the names and dates of death of the people who are remembered in the project, and the penultimate page of its book version (McQueen 2010), display the single statement that it is 'dedicated to all victims of the Iraq war'. But while the work is dedicated to all victims, it only provides a partial accounting. Beyond the dead service members who are represented are thousands of injured British troops, as well as British civilians who died in Iraq and thousands of American dead and injured.

Also absent are the Iraqi dead, for whom no precise accounting has been given and whose numbers vastly outweigh British and American losses. In focusing on British military losses, the work operates through rather than challenges a well-established national distribution of sensibility. If the war was wrong, this is registered here through a proper accounting for military, that is, national, losses; Iraqi losses are acknowledged implicitly but not given an account. In this way the work can be said to signal indirectly a constitutive absence not just in its own form, but in the wider geopolitics of the war. The phrasing also asserts the place of the service personnel as being among all victims of the war, and thus their equality with other people wrongly killed, injured and harmed by it; their deaths are not just to be remembered, but wrong. If the illegality of the war has not been crystallised in formal, judicial proceedings, *Queen and Country* is a work that at least records it as problematic.

The work also marks out 'the Iraq war' as an event, but in a way that differs from other renderings. Here, the event begins on 21 March 2003 with the death of Captain Philip Guy of the Royal Marines at age 29 and ends on 12 February 2009 with the death of Private Ryan Carl Wrathall aged 21. But the event is also extended by the work, which both continues to enact the war as if it is not yet properly concluded, and which points, in a spare and inadequate way, to much greater losses, which are much harder for a British public to feel or perceive.

The Baghdad Car, 2010

In September 2010, more than a year after the formal withdrawal of British forces from Iraq, an unusual object – the wreck of a car destroyed in the bombing of a market in Baghdad in March 2007 – appeared in the atrium of the Imperial War Museum (IWM) London, a few hundred yards south of the River Thames in the Borough of Southwark (Figure 4.4). The IWM had been established nearly a century before, while the First World War was still taking place, in order 'to collect and display material as a record of everyone's experiences during that war – civilian and military – and to commemorate the sacrifices of all sections of society' (Imperial War Museums 2017a). Initially called the National War Museum, the institution was renamed to take account of Britain's colonial 'dominions', and while today the IWM (which includes four other sites in London and elsewhere in the UK) describes its work as being 'to help people, as global citizens, to make sense of an increasingly unpredictable world' (Imperial War Museums 2017b)[15] its focus is on the experience of 'Britain, its former Empire and the Commonwealth' (Imperial War Museums 2017a). The IWM is thus another important institution that both expresses and shapes British self-understanding of empire and nationhood in relation to historic and contemporary conflicts.

The wrecked car was unusual because, at that time, the atrium – a huge, light space that was the visitor's first point of encounter with the museum – was otherwise

Figure 4.4 *Baghdad, 5th March 2007* in the atrium of the Imperial War Museum, London (2010). Photo by author, courtesy of Jeremy Deller.

taken up with the display of military vehicles and weapons (Ingram 2012, 2016). Indeed, the car was described at the time of its accession as being unique in the IWM collection, in that it was an object that had been destroyed in war. However, the car was not exactly a conventional museum exhibit either, in that its caption described its display as taking place in collaboration with Jeremy Deller, an artist well known for imaginative and politically astute public space interventions. Here, then, we pick up the thread of Deller's idea of displaying an object from Iraq in public space in Britain with the title of *The Spoils of War*, discussed in the Introduction, in a different setting and with a different title.

Installed at IWM London, the object spoke ambiguously on the one hand to the idea of the museum as a documentary institution geared to the presentation of artefacts, and on the other to its role as a space for artistic experimentation with how war is conducted, encountered and understood. The car had undergone several evental transformations. It had been destroyed in a massive, deadly explosion that took place on 5 March 2007 devastating the historic al Mutanabbi Street book market and surrounding cafes, with the loss of 38 lives and more than a hundred injuries. It had then been removed to a compound on the edge of

Baghdad. It was then transported to the Netherlands, where it was used in public protests and in an exhibition (Kluijver 2010), where it was given no caption or title. Having used the car as the focal point for a touring project in the US (Deller 2010; see also Sahakian 2012), the car was then accepted into the collection of the IWM, becoming in some sense a museological artefact. In the atrium of IWM London, however, the car was exhibited with the following title and caption:

> Baghdad, 5 March 2007
> A new display with Jeremy Deller

As if to reinforce a sense of connection between the explosion and the car, the work took the date on which it had been destroyed as its title (Figure 4.5). While the Museum and the artist insisted that the car was not an artwork, it inevitably became one in virtue of its association with Deller, and was received as such by a series of commentators (Ingram 2012, 2016). But what I would like to highlight here is something that went unremarked among the various responses to the object. While Deller's original proposal for the Fourth Plinth had gone under the title *The Spoils of War*, the car was presented at the IWM without this polemical context, as more befitting of a museum object; that is, as a 'display'. Nevertheless, extracted from the capital of a country under occupation by its former imperial power and held to constitute material evidence of an otherwise

Figure 4.5   *Baghdad Car* at the Imperial War Museum, London (2010). Photo by author, courtesy of Jeremy Deller.

obscure reality, its appearance at the museum must also be seen as echoing colonial archaeological practices, albeit again ironically, and thus as being in communication with the Event of colonialism itself.[16]

The appearance of car speaks to the genealogy of the museum as an institution that has been central to the production of knowledge and of the public sphere, as well as a sense of Britain's role and place in the world; a space in which certain kinds of events happen in order to make other events more comprehensible. Its appearance also therefore illustrates the process whereby museums participate in the aesthetic production of national identity as something that can be felt and touched (in this case, illicitly), as well as seen and heard, and the way that art both appears within, and may serve to complicate, the veridical (or truth-making) discourses that otherwise suffuse and structure the museum as institution.[17]

Conclusion

This chapter has explored how the event of the Iraq war has been refracted through museums and museum-mediated artworks in Britain and has sought to situate museums, exhibitions and museum-mediated artworks in terms of a genealogical account of both art and geopolitics. It has sought to demonstrate the necessity of situating artworks not just in relation to specific geopolitical events, or the *dispositif* of art at a general level, but in relation to the specific institutional contexts and spaces of museums themselves, while drawing out the ways in which artworks appropriate, question and point beyond them. As the chapter has explored, British museums have been implicated in coloniality and militarism and in the formation of European and national subjectivities according to particular kinds of hierarchies and categories, through and in relation to which they have classified objects, people, places and events, and which those objects, people, places and events, once classified, have helped to reify. As the chapter has also highlighted, the pressure of events has catalysed a turn towards more critical and reflexive artistic, curatorial and museological practices, questioning and challenging the ideological and discursive parameters of the museum, but the question remains of what a more fully decolonised museum might look like.

In focusing on the period from 2006 to 2008, the chapter also recovers a moment of crisis in the distribution of sensibility, when structures of feeling concerning the wrongness of the war crystallised, hastening the departure of Tony Blair from office. It is around this time that a range of actors associated with government and the military began to develop the idea that the proper relationship between the state, military and society had broken down and needed to be repaired, an idea that led to a range of new initiatives to bring the military more fully into national life and to promote a military ethos. The artworks discussed here endure as traces and reminders of a moment before such initiatives gained ground.

Notes

1   Al-Azzawi has become the most prominent Iraqi artist in the diaspora since leaving Iraq for London in 1976. His work is held in the British Museum and in Tate Modern as well as in many collections in the Middle East and elsewhere. Al-Azzawi's recent influence is reflected in his co-curation of the *Art in Iraq Today* project (Faruqi 2011). In 2016, a major retrospective of 500 works was held across two major galleries in Doha, Qatar. Several of Al-Azzawi's works evoke Picasso's *Guernica*: these include *Sabra and Shatila Massacre* (1982–1983), dealing with the massacre of Palestinians in a Lebanese refugee camp, *Mission of Destruction* (2004–2007), which addressed the looting and destruction of Iraq's cultural heritage in 2003, and *Elegy to My Trapped City* (2011), which responds to the destruction of Baghdad, the artist's home city. Al-Azzawi is sometimes likened to Picasso himself and has discussed his career and work in interviews (Harris 2016; Smith 2016). He has recounted his experience of authoritarianism in Iraq before his move into exile (see Chapter 5), and has forcefully critiqued recent Iraq National Pavilions at the Venice Biennale (see Chapter 8).

2   The list of studies referenced here is partial and selective: amid an extensive literature in museum studies and beyond I mention works that have been useful to me in thinking about the museum as a geopolitical institution and discourse. While many critiques of the museum have been informed by Foucauldian thinking, the studies I cite here and below also reflect feminist, postcolonial and decolonial challenges.

3   European explorers overlooked the 'fairly accurate' accounts given by Islamic geographers and historians of the locations of ancient cities (Bernhardsson 2005, p.35).

4   See http://www.rgs.org/NR/rdonlyres/6AD2A463-2F70-4242-AD4C-EC370E87E28D/28945/TheGoldMedalsPDF1.pdf. In 1838 Colonel Francis Rawdon Chesney received the Founders' Medal 'For valuable materials in comparative and physical geography in Syria, Mesopotamia and the delta of Susiana'; in 1839, the Patron's Medal was awarded to Dr Edward Rüppell 'For his travels and researches in Nubia, Arabia and Assyria'.

5   In contrast with the British Museum, the Iraq Museum established in 1923 was framed as being 'national' rather than 'universal' in orientation and was conceived initially as a repository of archeological finds rather than a teleological, ordering institution' (Bernhardsson 2005, pp.150–151).

6   The geographies of anti-war protests in Britain since September 2001 have been explored in Gillan *et al.* (2008) and Phillips (2009).

7   See https://www.forumauctions.co.uk/35145/Banksy-Pax?auction_no=2024&view=lot_detail (accessed 14 December 2017). Images of many of the works from this show, as well as numerous photos showing Gentleman's designs and variations on them are included in Stop the War Coalition (2011).

8   Though this was not highlighted in the exhibition itself, we can note its proximity in Oxford to the Ashmolean Museum, which holds an extensive collection of artefacts taken from Mesopotamia in the colonial period.

9   See full poll results at http://cdn.yougov.com/cumulus_uploads/document/raghpsamv0/YG-Archives-Pol-Trackers-Iraq-130313.pdf.

10 Surveys can also provide insight into how memory of the event changes over time. By 2013, more than 50 per cent of respondents held that the US and Britain had been wrong to take military action, with only 27 per cent stating it to be right (over the

course of ten years, 'don't knows' had doubled to 20 per cent). Research using the 2011 British Social Attitudes survey found that while 94 per cent of people expressed support for service personnel who had served in Iraq, 58 per cent of respondents (56 male; 60 female) agreed that 'The UK was wrong to go to war with Iraq in 2003' (Gribble *et al.* 2012, p.145). Research based on the same study found that the most widely held explanation for the war (at 47 per cent) was that it had been undertaken to ensure Western oil supplies (Gribble *et al.* 2015).

11   Laing's anger at the war is evident in correspondence (seen by the author) between him and the NAM regarding the exhibition of his work; also Farquhar Laing, personal communication.

12   Laing correspondence with NAM.

13   The 790 objects comprising the camp were acquired by the Museum of London following Haw's death and form part of its collection (Kavanagh 2015).

14   As the Museum announced, 'The artist has an open brief to make new work, which reflects the current situation in Iraq and is influenced by what he sees and experiences during a visit to that country. The timing of the visit is dependent upon a number of factors, but it is hoped that this will take place during the next two to three months' (Imperial War Museum 2003).

15   In its annual report for 2016–2017, the IWM describes its work in the following way: 'Our vision is to help people, as global citizens, make sense of an increasingly unpredictable world. We do this, in part, by helping people have a deeper understanding of the connections between past conflict and the contemporary world. This is about exploring the way war has shaped the local and the global, about appreciating diverse views, and about challenging our audiences to become ready to engage in difficult decisions for themselves, their communities and their world' Imperial War Museums (2017b, p.4).

16   As well as a public discussion of the car, the Museum also organised showings of films made by people of Iraqi heritage about the bombing and its impact on the Shabandar Café, a centre of literary and political discussion on Mutanabbi Street. Nevertheless, it is the car that forms the most enduring presence in the Museum.

17   The car appeared again in 2012 in a retrospective of Deller's career at the Hayward Gallery on London's South Bank along with volunteer experts and Iraqi people who were on hand to engage visitors in conversation, a format that echoed Deller's work with the car in the US (Deller 2012; for a telling critique of this setup, see Sahakian 2012).

Chapter Five
Iraq Beyond Iraq

In 2009, Zainab Bahrani and Nada Shabout curated the exhibition *Modernism and Iraq* at the Miriam and Ira D. Wallach Art Gallery of Columbia University. Their curatorial, analytical and political framing of this exhibition, which presented Iraqi art in relation to modernism in a way that challenged Western categories of aesthetics and art, enables us to think about Iraqi art history both genealogically and geopolitically and to appreciate the ways in which artists of Iraqi heritage experienced the 2003 war and responded to it. As Bahrani (2009, pp. 13–14) explained, with reference to Rancière's thinking on the politics of aesthetics,

> The exhibition *Modernism and Iraq* is a presentation of art, but it is also polemical in its intervention, because what we hope to address, to some extent at least, is the politics of perception that is central to the question of what Iraq means to the US public today.

The exhibition and this statement are also pertinent to the meaning of Iraq for Britain in light of the 2003 war. As this chapter will explore, drawing critically upon the work of Bahrani and Shabout in concert with other interventions in Iraqi politics and aesthetics further allows us to understand Iraqi art history in relation to questions of coloniality and decolonisation as well as the more recent political history of Iraq and the geopolitical events leading up to and beyond the 2003 invasion. These events include the rise of Baathist rule and the regime of

Geopolitics and the Event: Rethinking Britain's Iraq War Through Art, First Edition. Alan Ingram.
© 2019 Royal Geographical Society (with the Institute of British Geographers).
Published 2019 by John Wiley & Sons Ltd.

Saddam Hussein, the Iran-Iraq war, the 1991 war and the long decade of sanctions and bombing that followed it, as well as Saddam's repression of the Iraqi Kurds and other groups in Iraqi society, each of which in different ways flows through, and is counter-actualised in, Iraqi artists' responses to the 2003 war. Exploring this body of analytical, curatorial and artistic work also goes some way towards addressing the absent, brief or isolated treatment of work by artists of Iraqi heritage in much writing on the war in political geography and international relations.

Recognising the growing body of academic, curatorial, artistic and activist work that addresses and extends Iraq's complex and multifaceted art history within and beyond the country, this chapter highlights the experiences and works of artists of Iraqi heritage who have, or have come to develop, ties to Britain, and whose work has been created and exhibited here. This work can be regarded as being engaged in the unfolding of Britain's Iraq war no less than works by artists more commonly identified with British identities and institutions. In drawing out some of the complex and multifaceted events involved in a series of artworks, I highlight the ways in which they undo a topographic sense of 'here' and 'there', expressing and enacting topological spaces of connection and attachment.[1] The discussion therefore draws out how, while the violence of the war was concentrated in Iraq, it was also experienced and appropriated well beyond Iraq itself by people with connections to the country. Considering diverse works by diverse artists, whose pathways through and relations to the event are also diverse, further underscores its multiplicity.

After developing a genealogical understanding of Iraqi art history in relation to geopolitical events, the chapter considers works by five artists who encountered and addressed the war in distinct ways.[2] While the works discussed here all relate in some way to the ostensible event of the 2003 war, when considered individually and collectively they also question the extent to which we can understand it as a singular thing, rather as being made up of, and connected with, numerous other episodes of intense violence, each of which might be appropriated, reassembled and serialised in different ways.

Iraqi Art History and Geopolitical Events

As Zainab Bahrani (2009, p.14) writes, the middle of the twentieth century saw the emergence of an 'optimistic Modernist art in Iraq'. Three major art institutions – the College of Fine Art, Academy of Fine Art and Institute of Fine Art – had been created and several artist groups and schools had been formed. Having achieved formal independence, the government of Iraq sent promising artists abroad, to art schools in Europe and in China, and several, having returned, created new programmes and took important advisory positions in official institutions (Al-Gailani Werr, 2013). By the 1950s, and building on deep knowledge of the country's cultural heritage, an emerging generation of artists, 'profoundly secular and nationalistic' in orientation (Bahrani 2009, p.14), sought to forge not only a new form of identity that would draw together diverse ethnic, religious and regional affiliations, but new ways of thinking about and encountering identity through artworks. Though initially detached from most of the population, the work of this generation of artists became immensely significant for the political aesthetics of Iraqi statehood.

As Nada Shabout (2009) argues, the early- and mid-twentieth century Iraqi artists who studied in Europe should be regarded not as imitating modernism nor as offering an alternative version of it, but as equal participants.[3] As Bahrani (2009, p.15) explains, '[t]hese artists… thought of their work as Modern, as participating in Modernism, as experimental but at the same time as Iraqi or Arab'. Furthermore, if we understand modernism as being predicated upon the rejection and transcendence of mimetic representation, then it is also necessary to recognise that in the Middle East, 'representational mimetic art had never achieved the prominence, or the equation with art itself, that it had in the West' (Bahrani 2009, p.16). The aesthetic parameters of mid-twentieth century Iraqi modern art thus differed in some important respects from European variants. As Bahrani writes,

> abstraction and nonmimetic art forms are conventional and even conservative aspects of high art in Islamic artistic tradition. The rejection of figurative imagery in Modernist art in the West could not be relevant in the same way as in a place like Iraq. There, the identifiable characteristics of Modernism, as divergent from the representative system of art, naturally merge with the older modes of artistic production. (Bahrani 2009, pp.16–17)

As Bahrani (2009, p.17) continues, 'this kind of Iraqi Modernism (which no doubt exists elsewhere also) … undermines the category of Modernism itself'. Still further, while it is now widely accepted that 'European' modernism appropriated many non-European sources of learning and inspiration, it is necessary to note that certain key European modernist movements, including Surrealism, as well as artists including Charles Sargeant Jagger, Henry Moore and Giacometti, were inspired specifically by ancient Near Eastern artworks. European modernism thus contained traces of ancient Mesopotamian cultures that had themselves developed highly sophisticated forms of art and aesthetics (Bahrani 2014). Iraqi and European cultures are more entangled than is often recognised.

The most significant expression of mid-twentieth-century Iraqi modernism is the Freedom Monument (Hasb al-Hurriyah) installed in Central Baghdad following the revolution of 1958.[4] As Kanan Makiya wrote (under the pseudonym Samir Al-Khalil), the Monument 'belongs to the revolutionary city of the late 1950s and 1960s' (Makiya 1991, p.82).[5] The Monument was commissioned by the new republican government to celebrate the revolution, an anticolonial event in which King Faisal II was killed and the bronze statue of General Maude, who had claimed Baghdad for the British in 1917 (Gregory 2004a), was torn down.

The monument was designed by Jawad Selim, a central figure in Iraq's modern art history, and Makiya (1991, p.81; also Jabra 1961) calls it 'probably the most important work ever commissioned from a modern Arab artist'. Selim had received his initial training in Baghdad and was then sent on government scholarships to study in Paris (1938–1939), Rome (1939–1940), and at the Slade in London (1946–1949). He was a founding member of the Institute of Fine Arts (1949) and, having been a member of the influential Pioneer group of artists, founded the Baghdad Group for Modern Art in 1951. In a public lecture in the same year, Selim had lamented the poor taste of everyday Iraqis and noted that artists were widely considered to be enemies of the people (Makiya 1991), just as he was on the brink of international recognition. In 1952, Selim entered a competition organised by the Institute of Contemporary Arts in London to design a monument to the unknown political prisoner; while he did not win, Selim's proposal was among a few dozen selected from thousands of entries to be exhibited at the Tate (Makiya 1991). Selim died in 1961 while overseeing the installation of the Monument in Revolution Square in Baghdad, where it continues to stand.

The Monument (Figure 5.1) is organised around a sequence of figures and scenes that recall ancient Assyrian reliefs and cylinder scrolls, and which are read, like Arabic script, from right to left. The arrangement of the figures signifies a temporal progression, from the pre-revolutionary period of oppression and

Figure 5.1   Freedom Monument, Baghdad. Source: Abdullah Alanzy, https://commons.wikimedia.org/wiki/File:20160102-Tahrir_square_Baghdad.jpg. Licensed under CC-BY 4.0. Image cropped and anonymised by the author.

resistance, through the event of the revolution in the centre, to an optimistic post-revolutionary future on the left. In the central panel, there is a depiction of a political prisoner, and of a male soldier as a liberating force, breaking forcefully through metal gates.

As Bahrani and Shabout (2009, p.98) write, this work 'evokes a shared humanity and defines a sense of historical identity and national character' and they note how it 'inspired virtually an entire generation of younger artists'. While the Monument was subsequently appropriated by authoritarian nationalist discourses, we can, if we interpret it in relation to modernism, also read it as anticipating, resisting and criticising such appropriation. If Makiya (1991) is right that the monument was raised up off the ground against the wishes of Selim, who wanted it to be accessible to people at street level, then this might further lead us to see it as a metaphor for freedom yet to be realised. The Monument thus embodies a complex relationship to events. Its construction and unveiling were events; it is 'about' an event, and it has continued to act in relation to ongoing events, in the recent past becoming a focus during the Arab Spring for protests against corruption and misrule, while continuing to posit possible futures. As with the figure of the *lamassu* and Al-Azzawi's *Blessed Tigris*, and as this chapter considers, ancient objects, forms, motifs and themes continue to be reworked and appropriated in artworks that respond to ongoing events in Iraq, enacting possible pasts and futures.

Increasing political direction of political and cultural life became evident after the revolution, and intensified with the Baathist coup of 1963. As Al-Azzawi has recalled of this event, 'There was an atmosphere of fear... If anyone thought you were a Communist, they would come and pick you up' (in Smith 2016). At the same time, buoyed by growing oil wealth, Iraq came to assume a leading position in the emerging Arab art world. In 1974 Baghdad hosted the inaugural Arab Art Biennial and Iraqi artists participated in art networks emerging between non-aligned countries of the Global South (Gardner and Green 2013).[6] Nevertheless, the numbers of people leaving Iraq increased again after the second Baath coup in 1968 and again after 1979, as Saddam Hussein moved to repress perceived political enemies, and as a result of the Iran-Iraq war and the brutal campaigns against the Iraqi Kurds. Many artists were among those who left, some not returning from scholarships abroad in neighbouring countries, the Soviet bloc, Western Europe and North America. For some Iraqis fleeing the regime, Britain, while open to them, was not a comfortable place during the 1980s, when the Thatcher government was providing material support for Saddam's regime. Not only did British companies support Iraq's weapons programmes, however; the Morris Singer foundry in Basingstoke was the only facility capable of forging the giant hands (modelled on Saddam's) used in the huge, militaristic *Victory Monument* that was built in Baghdad after the Iran-Iraq war (Makiya 1991; Al-Gailani Werr 2013). Several of the Pioneer generation of artists and members of the Baghdad Group, meanwhile, were drawn into designing monumental works and structures glorifying Baathist rule and Saddam himself, while the regime selectively appropriated ancient and Islamic symbols, sites and objects, presenting Saddam as the culminating figure of Iraqi history (Davis 2005; Al-Gailani Werr 2013).

As a number of writers have argued (e.g. Davis 2005; Shabout 2007; Al-Ali 2011), it is a mistake to conflate Baathist rule and the regime of Saddam Hussein with Iraqi society, or to see all art and cultural and intellectual production that took place under the Baath as somehow complicit in its rule. For one thing, issues of consent and autonomy are not straightforward under authoritarian rule, and the absence of open opposition or visible resistance, and even compliance and accommodation, cannot necessarily be equated with active support (Scott 1990). As Eric Davis has argued with particular reference to Iraq,

> Even the under most oppressive regimes, there is always ongoing resistance by subaltern groups, even if it does not entail direct challenges to the regime's prerogatives. Moving beyond force and coercion to the struggle over history and culture reveals a contested arena in which symbols, metaphors, allusions, and double entendres belie a rich underlying debate that encompasses myriad questions concerning social equity, cultural pluralism, and political participation. (2005, p.17)

The National Museum and the Museum of Modern Art (named the Saddam Arts Centre during his rule) continued to function as repositories of ancient, traditional and modern art and artefacts, and as sources of inspiration and indeed resistance to political events, even as they were caught up in ideological imperatives and political manipulation.[7] While the Baathist state funded and sought to direct cultural production, and while many artists produced work glorifying the regime, it cannot be assumed that all art produced during this time, or the artists who produced it, were complicit in the regime's authoritarianism, or indeed that they necessarily knew its full dimensions.

Nevertheless, many individuals and groups did also oppose the rise of the Baath and Saddam, and were dismissed from their jobs or had their careers obstructed, or were killed, tortured and driven into exile as a result. It is also necessary to recognise the particular way in which the Kurdish regions of Iraq were targeted by systematic mass violence and that Kurdish movements not only resisted this and asserted distinctive identities, but have pushed for autonomy and independence from Baghdad, even while such resistance has entailed internecine struggles and been accompanied by gender-based power and violence. If national identity and historical memory remain essentially contested (Davis 2005), this is in large measure because Iraqis, while often expressing strong support for Iraq as a polity and dismay at the course of the war and politics since 2003, and while often maintaining and developing complex, plural identities and affiliations that do not fit easily in to stereotypical or received schemas (Al-Ali and Al-Najjar 2012), have fundamentally different experiences and interpretations of defining events in the country's geopolitical formation and political history (see also Batatu 1978). While attempts to assert and impose a singular vision of national and state identity are always fraught with risks of conflict, authoritarianism and fragmentation, these risks have been actualised many times over in Iraq, and the resulting political tensions and cleavages continue to shape the politics of the country within its borders and beyond.

While not downplaying the forceful and indeed existential nature of these questions for people of Iraqi heritage within and beyond the country, and while not assuming that all Iraqis have been equally involved in, or equally affected by, political events, it is not the place or goal of this book to make political or ethical judgements about the relationships of art and artists to the former regime or to the 2003 invasion, which some people in the diaspora and in Iraq supported.[8] Rather, my discussion aims to contextualise the complex, multilayered experiences and works of artists of Iraqi heritage who have, or have developed, connections with Britain and whose work has been created or exhibited here.

Here it is relevant to expand briefly on how work by artists claiming an Iraqi Kurdish identity often bears a distinct relationship to the politics and geopolitics of Iraqi state formation and to questions of political violence. In this regard, politically-oriented work by Iraqi Kurdish artists who have left Iraq and settled in Britain has often tended to address the violence of the Saddam regime towards Kurdish communities and themes of Kurdish resistance, displacement and identity rather than the 2003 war *per se*. The evocative *Displaced* series of drawings of Osman Ahmed, which is informed by Goya's *Disasters of War* and Picasso's *Guernica*, and which is held by the Imperial War Museum (IWM) London, focuses on the experience of Kurdish people during the Saddam regime's *Anfal* campaign, which is understood by the artist as a genocide and to which the work is understood as bearing witness (Ahmed 2007). The work of Walid Siti (see also Chapter Eight), meanwhile, has often focused on the materially and symbolically important figure of the mountain for Kurdish identity and resistance, while other works have dealt with the politics of water in ways that might be read as implicitly going beyond or calling into question the geographical and political parameters of the Iraqi state. While I later discuss other works by artists of Iraqi Kurdish heritage and artworks and events taking place in Iraqi Kurdistan, an effect of focusing on the 2003 war is that such works receive less attention than they otherwise might.

In the following sections, I discuss works created by five artists with diverse experiences of Iraqi political and art history and the 2003 war. For the title of each section I use not the name of the artist but of a work by them that expresses a particularly geographical relation to events. While the first of these projects explores ways in which people in Britain might have come to know about Iraq or not during the 2003 war, the others embody Iraqi experience and aesthetics in a more direct way.

Small Medium Large | Man Woman Child

The issue of the complexity and multiplicity of Iraqi and British identities is evoked in the work *Small Medium Large | Man Woman Child* (2009), created by Rabab Ghazoul for the National Eisteddfod of Wales, as is the question of how, and how much, people in Britain might know about the country and ongoing events there. Rather than trying to convey a sense of the reality of Iraq or the war with affective force, this work sought to prompt participants to become more aware of their own *lack* of experience or knowledge and of how, especially in times of war, narratives of other places and people become flattened, narrowed and militarised (interview with author). While it cannot be reduced to a biographical reading, *Small Medium Large | Man Woman Child* emerged to some extent from how the war affected Ghazoul in terms of her connections with Iraq. As stated on the artist's website (Ghazoul 2017a): 'I was born in Mosul, Iraq. I've lived and worked in Cardiff, in Wales since 1993.' As Ghazoul has also related: 'I don't term myself An Iraqi Artist, well I am and I'm not but I'm more not than am'; Ghazoul has also mentioned being involved in campaigning against the sanctions placed on Iraq from 1991 to 2003 (interview with author).

The project was presented at the Eisteddfod (a cultural festival dating back to 1176 and revived in 1861; Eisteddfod 2018) as a market stall, on which were arranged a series of T-shirts bearing a slogan linking the form 'I ♥…' or, in Welsh, 'DW'IN ♥…' with the name of a place in Iraq, such as Karbala, Diwaniyah or Basra (Figure 5.2).

Figure 5.2 *Small Medium Large | Man Woman Child*, Rabab Ghazoul (2009). Image courtesy of the artist.

Displayed on other T-shirts, which could be purchased, were texts composed by project participants, who had been invited to follow a set of instructions:

> Select an Iraqi city or town from the list provided
> Try to discover something about this place
> Submit a paragraph detailing your experience of the task
> Include, somewhere within your response, a reference to Wales
> (Ghazoul 2017b)

The project was informed, first, by an awareness of the Eisteddfod's concerns with place and identity:

> I was thinking about the context of the Eisteddfod as a celebration of the Welsh language and there's lots of stalls and a lot of t-shirt stalls, and I wanted to replicate that, and was thinking about the 'I heart New York' t-shirt, because there's something about that, that it conveys, 'I have been to this place and I have knowledge of it'. It's a kind of shorthand for it. (interview with author)

This was relevant in the context of British understanding of Iraq precisely because most British people would not be able to say the same of that country: 'we can't go, you can't have that knowledge, you can't say that' (interview with author). *Small Medium Large…* thus raised the question of the limited or absent knowledge people might have of Iraq, a country in which British military forces were deployed, in the context of an event intensely concerned with national identity. The project was also informed by a concern with the reductive and militarised nature of narratives concerning Iraq:

> we now have a whole set of knee-jerk associations with 'Iraqi' – even the places, Basra, Fallujah, Baghdad, Mosul, they immediately have certain very narrow connotations and it becomes a war-saturated narrative. (interview with author)

There was also a personal dimension to this line of thinking:

> It's the unconscious cultural and historical references that mean you're from a place, you're British and European because those are the unconscious references. And I'm aware that I don't have those in terms of Iraq… I vaguely know, that because I come from the north of Iraq, when I was growing up, I was told, 'Oh, you've got a really strong Moslani Iraqi accent', [but] I don't really know what that means. I remember asking my dad, 'What's a Baghdadi accent? What sort of accent is from Basra?' where of course I know what a northern accent is or a Scottish accent is. Those things really place you and when you are no longer somewhere you don't have all of that unconscious layering which makes you feel, in inverted commas, legitimate. (interview with author)

This was further linked to emotions experienced by Ghazoul around the time the project was conceived:

...this is going to sound really dramatic but at a certain point in my life I really mourned the part of me that was connected to Iraq and the Middle East and was only going to get weaker.... And if it ever did get stronger it would be going back as a Westerner... accepting that if you go back you're going to be a kind of tourist in the place of your birth. Which is fine. Actually I think we're all tourists. I'm perfectly happy to be a tourist. But I think at that time there was a sense of wanting to hang onto that and knowing I can't. (interview with author)

The texts produced by the project reflected the results of people trying to find out about places which had appeared in media coverage of extreme events during the war, but of which they had no direct knowledge. As one participant, who had selected Ramadi, wrote:

I wanted to discover something about this place that had no reference to war, but in the fifteen minutes I've spent searching I haven't managed it. That is quite sobering, and I feel a sadness for the city because of it. (Ghazoul 2017b)

As another wrote, having selected Balad,

I chose it because it sounds like a song when you read it out loud. I googled it. I found out that at the airbase there is a full size swimming pool. Cardiff has an Olympic size pool. The boast of the airbase was the pool. Cardiff boasts about the size of its pool too... What struck me was that in the 5 or 6 sites I looked for at Balad, all the information was related to the kind of Muslim who inhabited it and how much destruction had occurred there. As well as fact after fact about the army/ airforce and the military campaigns that had been launched from there. I was hoping to find out about famous inhabitants or see a resident's face in a photo. None to be seen. (Ghazoul 2017b)

From these responses it appears that what *Small Medium Large | Man Woman Child* elicited is not just an awareness of a lack of knowledge, but a sense of the difficulty of learning anything much at all about Iraq or Iraqi people, or the 'nightmare' (interview with author) they were encountering. As Ghazoul further relates,

I think this isn't maybe a political piece of work, although it encompasses things that inevitably are about power and knowledge and stories and a narrative and what is known and what is not known, and how do you shake up those spaces which have become quite immovable and fixed, and start to create a little bit of movement. Which is always there through people's curiosity and their desire to find something out; that creates an interesting space. (interview with author)

Small Medium Large was a participatory work, which enrolled people (Dixon 2015) in thinking about and producing, in effect, a small piece of geopolitical writing, reflecting on the limits of their own knowledge and how it had been affected by war and militarisation. Rather than a grand political gesture about the politics of the war, it worked at the level of the everyday, eliciting a degree of reflexivity as to its participants' positionality and locatedness with regard to the ongoing event, and an imaginative exercise in thinking about what was not known. The use of postcards, T-shirts and the well-known popular cultural form 'I ♥…', meanwhile, introduced a certain distance from ways in which the war was commonly mediatised. Rather than trying to remedy a lack of knowledge or in some way make the nightmare of the war present, the work allowed participants to experiment with their apparent lack of connection to 'war at a distance' (Kaplan 2013). In the following sections, I consider works that seek to enact much more direct encounters with the event.

My Country Map

Hanaa Malallah was trained by members of the Pioneer generation and in particular Shaker Hassan Al-Said, who had been taught by Jawad Selim and studied in Paris as well as Baghdad and who had, with Selim, founded the Baghdad Modern Art Group. In 1971, Al-Said founded the One Dimension group, which explored the multiple meanings of Arabic script via modernist techniques (Porter 2006, p.139). As Shabout describes, during this period Al-Said's work 'underwent a profound transformation' in which the artist 'specifically rejected representation as an aesthetic loaded with social and class significance' (2007, p.16). Al-Said's later work would often be characterised by geometric lines, dots and grids and script or script-like figures, layered over and through washes of colours of varying intensity.

While Al-Said had been able to develop his work during the 1970s and 1980s by participating in international biennials, the possibilities for artists who had remained in Iraq were sharply curtailed, first by the 1991 war and then by the isolation and hardship imposed by sanctions, as a result of which artists were largely unable to travel (beyond occasional trips to Amman in Jordan), or to access publications relaying international art world developments, or to obtain art materials.[9] While the ability of the government to exercise control was somewhat reduced and a small private market arose in Baghdad, the scope for art and artists was seriously constrained. During the 1990s, Malallah's work responded to the aftermath of the 1990–1991 war and sanctions, working with distressed, rough and found materials (Malallah 2010), while drawing on imagery from the Assyrian, Babylonian and Abbasid periods as well as theories and techniques developed by Al-Said and previous Iraqi modern artists. This work also responded to the closure of the Archaeological Museum, an important resource and reference point for artists in Iraq, as a result of the 1990–1991 war.[10]

In 2006, Malallah was forced to leave Iraq under threat from militias which, following the invasion, embarked on a campaign of violence against professional people, intellectuals and artists.[11] Important to Malallah's work following the invasion and during her move into the diaspora was what she came to call a 'ruins technique', which 'includes the burning, distressing and obliterating of material' (Malallah 2010), a process in which the work also becomes a ruin (Al-Baholy, in Malallah 2009). She has described this practice in an embodied, evental manner as 'writing by knife, painting by fire'.[12] This drew on her experience of Iraq during the 1990–1991 war and under sanctions as well as the 2003 war. As Malallah (2010) has written,

> To physically taste war is completely different than to experience it second-hand. The first lesson taught by physically tasting war is that ruination is the essence of all being: Death has no meaning and anything solid can be reduced to nothing in seconds. The learning of this process of vanishing, this morphing of matter to dust, of something into nothing, has led me to conclude that ruination, or destruction is hidden de facto in the phenomenon of figuration. Thus, for the last five years [my work has] explored the space located between figuration and abstraction, between existing and vanishing, a concept which for me also holds deep spiritual meaning.

The direct experience of war is thus central to Malallah's work, and by using processes that degrade and transform the materials with which she is working, she aims 'to engender the visceral experience of the reality of war irrespective of its geographic/political particular[s]' (Malallah 2010). In Malallah's work the events of war, embodied experience and art work are enacted together and flow through each other.

This approach is evident in a series of works created from 2003 onwards, which have been exhibited in Britain, Europe and North America as well as in recently created art institutions the Middle East.[13] In these mixed-media, canvas-based works, the city and the state are approached through techniques, materials and symbols that, as well as evoking the experience of war and having wider spiritual references, are associated with urban form, national identity and military organisations (see also Gregory 2010a). In earlier work, Malallah had addressed the experience of being in the city in works such as *Codes and Signs* (1998); from 2003 her work addressed the effects and experience of the invasion and occupation. Malallah has since developed a diverse set of practices, but her experience of war, survival and exile remains a central thread connecting them. I will explore this thread through two works made by Malallah following her move into the diaspora, *My Country Map* (2008) and *My Night 16–17.01.1991* (2012–2013), both of which have been exhibited several times in Britain and internationally.[14]

My Country Map (Figure 5.3) can be read as part of a series of works created immediately following Malallah's move into the diaspora and which includes

Figure 5.3   *My Country Map*, Hanaa Malallah (2008). Photo by author, courtesy of the artist.

Baghdad City Map, which incorporates military figures into a distressed carto-graphic rendering of the urban environment, and *Manuscript from Mutanabi Street*, which relates to the book market bombed in March 2007. Both of these works address the deeply embodied way in which the destruction of the city was experienced.

My Country Map, which measures two metres square, is comprised of layers of scorched and burnt canvas treated with oil paint and pigments. If this object might once have been a map of the country and its region, it has been violently trans-formed by tearing, burning and forceful rearrangement. If, as Kitchin and Dodge (2007) argue (also Gregory 2010b), maps can be regarded as events, then this work posits a map that has undergone an intense material-energetic transforma-tion, leaving it in tatters. The layers of canvas, with burnt edges and scorch marks, also recall the remnants of a shroud, while a handful of place names ('The Gulf', 'Baghdad', 'Mosul') endure in roughly the places one might expect to find them and the word 'Iraq' is burned into an area of canvas at the lower left edge. Patches of red and green, along with dark burn marks and faded white canvas, evoke Iraq's flag. If maps and flags typically serve as part of the ordering of territory and are reflective of governmental and epistemological authority, they can also exert strong affective forces and elicit feelings of attachment. Here the process of distressing, tearing and rearranging map and flag both expresses and parallels the artist's experience and feeling of what war does to nation, state and homeland. The work registers the actual disruption of the world, sensation and ways of making sense, while hinting at the virtual forces that have impelled these transformations.

As with the work of many other artists connected with the country, Malallah situates the 2003 war in relation to the more general geopolitical situation of Iraq and to other events, both punctual and extended, encompassing longer time spans than the post-2003 period. In particular, Malallah relates the 1990–1991 war as a pivotal, traumatic event that devastated Baghdad and the period of sanctions until the beginning of the oil-for-food programme as a period of particular hardship, during which cultural and public life was severely curtailed (interview with author).

Another of Malallah's works made in the diaspora, a large format (1.5m × 6m) mixed media and soft sculpture work (described as a 'polyptych': Park Gallery 2017) entitled *My Night 16–17.01.1991*, relates back to the bombing of Baghdad at the beginning of Operation Desert Storm. As Malallah relates, in an expression of how the event of war not only disrupts space and time, but ways of sensing and making sense, 'I was in Baghdad and the whole city just collapsed... I was just in shock... you think, "well, this is mad"' (interview with author). Malallah further relates that due to the government's control of the media, she did not see any images of the 1990–1991 war until she had left Iraq. She further emphasises the contrast between how the initial bombing was represented in Western media (in particular with reference to the notorious description of the bombing by a US reporter as being akin to a Fourth of July fireworks display) and the experience of surviving it.

The events to which *My Night 16–17.01.1991* relates are figured as in some sense continuous with the 2003 war through the use of techniques, materials and forms employed in earlier works, while also expressing their development. The work is much larger than Malallah's previous work; like several other Iraqi artists, Malallah has expanded the scale of some of her works in an effort to convey the enormity of the events to which they relate. Viewed from a distance of several metres, *My Night*, like many of Malallah's works, also resembles a devastated and traumatised landscape, viewed from above. A vaguely discernible, grid-like form might underlie the pockmarked, torn, burnt and twisted surfaces, which in several places appear to have been torn through, revealing a blood-red surface underneath. The finely detailed workings invite an examination from much closer up, at which distance the work cannot be grasped as a whole. At this range, it becomes apparent that bird feathers are distributed around the work and that a small hoopoe bird is itself incorporated within it, serving as messenger and symbol of survival.[15]

My Burning City

Suad Al-Attar settled in Britain after leaving Iraq in 1976, having received her art training and begun her career there in the 1960s. Until the 1990s, Al-Attar's work was known for richly detailed paintings that, inspired by Sumerian and Assyrian culture, depict mythical animals, romantic human figures and fantastical, verdant scenes of gardens and cities. Al-Attar's work has also been described as an expression of her relationship to Iraq and Baghdad in particular, which has also often appeared in her works.[16] As Daniel Robbins (2006, p.6), who has curated her work, has written,

> the perpetual sense of longing for 'home' has always been balanced by an awareness of the freedom that comes with distance. This freedom – a condition that gained added significance following the regime's rise to power under Saddam Hussein in the late 1970s – has enabled her to explore her relationship with her homeland and to develop a personal visual language with which to express it.

Al-Attar's work has changed dramatically in the course of events since 1990, to which her work since 2003 also refers. Al-Attar stopped painting altogether for a period after the 1991 war, turning instead to making small-scale drawings in ballpoint pen, accompanied by fragments of poems, (Robbins 2006, p.7), a move that to some extent paralleled the turn towards the private practice of constructing artistic books (*dafatir* – Shabout 2007) that developed among artists in Iraq and the diaspora in the 1990s.

In June 1993, Suad's sister Layla was killed along with her husband (a senior Iraqi government official) and several other people in a US cruise missile strike on Baghdad. Layla Al-Attar was herself a celebrated artist, who had trained at the Academy of Fine Arts in Baghdad and had exhibited at the first and second Arab Art Biennials (Barjeel Art Foundation 2017). Her work had been exhibited internationally, she had served as director of the Iraq National Museum of Art, and she was well known throughout the Arab world as a symbol of resistance to imperialism. The missile strike in which she died had been ordered by Bill Clinton, in response to an alleged car bomb plot ordered by Saddam Hussein aimed at killing then-President George H.W. Bush during a visit to Kuwait. In their attack on Baghdad, US warships fired more than 20 Tomahawk missiles at the compound of the Iraqi Intelligence Service, which Clinton stated had orchestrated the alleged assassination. Three missiles hit residential buildings adjacent to the intelligence headquarters in the Mansour district of the city, including Suad's home, in which Layla and her husband were living, their own house having been damaged in 1991 (interview with author).

The shock of this event was further expressed in Suad Al-Attar's work. As Robbins (2006, p.7) describes, '[t]he paradise pictures faded from her repertoire and the palette of brilliant colours began to ebb away, to be replaced by more subdued, muted tones'. As Al-Attar relates, in 1991 I could not paint at all for a whole year. And then, from 1993, really it's no colour. I just could not paint with colour. I look at the white canvas, and all that I could see is the mud, the brown, the earth and faint lines. (interview with author)[17]

When she returned to painting in the later 1990s, Al-Attar's work depicted what Robbins (2006, p.7) describes as 'monumental winged figures set against dark, brooding backgrounds, their faces lit by an unearthly, transcendent light'.

Al-Attar's work altered dramatically again with the beginning of the 2003 war. At this time, the artist was making 'spontaneous, direct' drawings every day but also writing a diary on the obverse of each page, finding it easier to express her thoughts and reactions in writing than in her usual art practice (interview with author). As she has related of these diary entries,

> [i]t's like I'm saying goodbye to the lands of my great grandfathers, I am saying good-bye to the Tigris and to the Euphrates and to Ishtar and to Gilgamesh and to Baghdad and to Basra and to Kerbala, to Nasiriyah, to Hilla… (interview with author)[18]

Figure 5.4 displays one of these sketches, which has never been exhibited, but which channels the artist's embodied as well as perceptual and cognitive experience

Figure 5.4   Untitled drawing, Suad Al-Attar (2003). Photo by author, courtesy of the artist.

of the war and response to it through techniques of drawing and through symbols drawn from ancient Mesopotamian cultures, which have been such an important resource and reference point for formations of Iraqi identity and experience. In the lower right-centre of the drawing, a male human face, apparently grieving and in pain, seems to be being encroached upon by two hooded figures, one in the light and one a dark shadow. Ranged against this scene from the left are a series of magical figures of human-animal form; there is a sense of tension and opposition between the two sides of the sketch as ancient and contemporary figures and forces seem to vie with each other. Here we see how drawing enacts a moment in which traumatic, deeply layered geopolitical events are transmuted into graphic form, both expressing and memorialising their affecting qualities.

The onset of war did lead to the creation over a period of months of a large-scale (1.8m × 3m) oil on canvas diptych, *My Burning City* (2003), which is also devoid of the greens and blues characteristic of her earlier work, and which is concerned with the destruction of Baghdad.[19] In this work, the canvas is dominated by dark, fiery reds, oranges and blacks, which seem to be consuming buildings, the outlines of which are only faintly visible; the scale of the conflagration overwhelms the city.[20] This work was exhibited at the Leighton House Museum in the Holland Park area of London in 2006, along with a series of works prompted by newspaper photographs of the bodies of dead and injured adults and children, painted with dark, muted colours and suffused by pain and horror. These works were presented in terms of the experience of destruction and suffering, rather than a directly stated political orientation to the war.

Black Rain

The paintings and mixed-media works of Yousif Naser, a former member of the Iraqi Communist party from Basra who left Iraq to escape Baathist rule and who worked with the Palestinian Liberation Organisation (PLO) in Lebanon before subsequently settling in Britain, address the 2003 war in light of European as well as Iraqi history. Naser has addressed the invasion in *Black Rain*, an ongoing series of works begun in the run up to the beginning of the war that deal with his response to the invasion, the occupation and their consequences for the country. Here I again describe how this project anticipated and emerged in relation to the invasion, and discuss its relation to multiple other geopolitical and art-historical events.

The *Black Rain* project arose collectively, out of a workshop with other artist friends in the build up to the war, which did not come as a surprise. As Naser recalls, 'I felt the war was coming; I was not under any illusions, I was sure the war was coming. We had to do something' (interview with artist). While the project was initiated with other artists, writers and poets, and exhibited in this form in Loughborough and Sheffield, it is Naser who has continued it as a series of works under the same title.[21] These large format

Figure 5.5   *Black Rain*, Yousif Naser (2010). Image courtesy of the artist.

works on canvas (up to 3m × 5m; Figure 5.5) are to a significant extent inspired by the German Expressionist painters of the early twentieth century, who anticipated, experienced and responded to the First World War. Indeed, Naser studied for a time in Dresden with artists who had been taught by some of the original members of the Expressionist movement (interview with author). Expressionist artists sought not to represent an external world but to paint the internal anxieties aroused by modernity, and the bold, dark forms and sense of movement in the *Black Rain* series recall such works as George Grosz's *Nudo* (1913), *Aerial Attack* (1915) and *Explosion* (1917). It is worth noting that the political implications of Expressionism subsequently became the focus of a seminal debate among Marxist theorists (Adorno *et al.* 2007). While Georg Lukács was dismissive of Expressionist and Dada experiments with fragmentation and violent, forceful affects, Ernst Bloch (in Adorno *et al.* 2007, pp.16–27) argued that Expressionism was concerned with authentic expression and a search for the human. As Naser has stated, the ways in which the Expressionists (and by extension Dada) worked through the traumas and implications of the First World War and anticipated the disasters to come remains relevant to the modern political history of Iraq (interview with author).

As in a Dada collage, the *Black Rain* works also sometimes include fragments of newspaper with stories related and unrelated to the war, blending horrific events with the banality of normal life removed from the warzone. The *Black Rain* works are dominated, however, by lines, blocks and sweeps of black paint suggestive of cloud, which sometimes surround spectral stylised figures suggesting human bodies, fish, buildings and aeroplanes as well as abstract forms, all caught up in a chaotic maelstrom. If the war was not a surprise, the *Black Rain* works register the manner in which it has been encountered as a profound affective disturbance, without offering any kind of transcendence.

The approach used to create the works also recalls modernist techniques. Naser took steps to bring the war into his studio via representations and materials as he created the works:

> War invaded my studio at this point. War images started to grow on the walls. News flashes blared out of the two radios I had placed strategically in my studio... The phone rang incessantly and friends streamed in and out... The usual paintings on the wall of my studio were removed along with the chairs. I replaced the paintings with large sheets of paper all the way round the four walls... inserting in between them all sorts of printed material about the war: newspaper cuttings, maps, photographs... Pictures depicting scenes of explosions, destruction, planes, war helmets, rifles, smoke, sand stained by the leaking tanks (I scattered sand on in my studio entrance and dripped oil on it...). (Naser, undated exhibition catalogue)

Black Rain cannot only be understood as channelling European modernist art techniques, however. This series also adopts and reworks ancient Mesopotamian religious and mythical motifs, which become a way of encountering and seeking to make sense of the imminent geopolitical event:

> We are the nation that invented the flood; the great flood is in our mythology. The Old Testament took it from Gilgamesh and Gilgamesh took it from somewhere. So the flood used to come to cleanse the gills of human beings and the dirt of the earth. So we need the great flood. I thought maybe this war will be the second or third great flood that cleanses everything and then Iraq starts again. But this will be with a very heavy price. It's the same as in the old Gilgamesh epic when the gods decided to flood the Earth, they left only one person, only Utanapishtim, to live and they killed the rest. It's the same now; I thought it will be very, very bloody. ... Now this will happen in our lifetime; it's next week for god's sake... (interview with author)[22]

While the *Black Rain* series includes discernible figures, however, Naser, in keeping with Expressionist tenets, does not understand these works as representational or illustrative:

> People have expectations when you paint about war, they expect to see mutilated bodies and bombs and destruction. Or illustrations like that. And that will reduce you

as an artist, it will put you in a corner. ... There are no bodies at all in my paintings, there is no one hung to the roof or whatever. And there are no bombs or anything. I paint about paint. It's not about the actual battlefield. (interview with author)

This approach further relates to Naser's experience of the civil war in Lebanon after leaving Iraq, at which time he worked for the propaganda department of the PLO.[23] As he relates, referring further to existentialist thought:

I was there, I felt it, I know what it means when you think you are going to die in a minute. I cannot describe that now. So that is what I paint about. ... To look into pain and wounds in a very realistic way, to see it and feel it and touch it. There is nothing else, only yourself. There is no escape. So that is what I paint about. ... I am not trying to please anybody and I am not doing therapeutic work. ... What I try to do is not to paint the people shouting, but the scream, the pain. (interview with author)

The *Black Rain* series is not just a direct 'response' to the 2003 war, then, but an eventual assemblage that contains distinct 'sites of time' (Lyotard 1992, in Pollock 2013a, p.xxvii). Painting here works as a way to materialise the inexpressible experience of both imminent mortal danger encountered during war, and the pain that it causes, while alluding to their virtual character. As well as German Expressionism and Mesopotamian cultures, then, there are also art-historical resonances with the work of Francis Bacon, which was strongly informed by Nietzsche's account of subjectivity and by existentialism (Hammer 2013). As Deleuze recounts, Bacon stated that he sought 'to paint the scream more than the horror' (Bacon cited in Deleuze, 2005, pp.42–43). In materialising and thus externalising and objectifying these violent affects, and while not representing a move beyond them, Naser's work might nevertheless also be read as engaging in resistance to violence and death.[24]

Buhriz

Having grown up in the city of Buhriz, some 25 miles north of Baghdad, Satta Hashem left Iraq in 1978 in order to 'escape' the regime and what he describes as its 'aggressive chauvinism and dictatorship' (Hashem 2007a), which he had opposed. Hashem received art training in Algeria and the Soviet Union and spent time in Iraqi Kurdistan, before moving to Sweden and then to Britain. Settling in Leicester, Hashem has been involved in a series of public artworks and projects working with refugee artists while developing his own work. Here I will discuss his drawing practice and then a series of paintings created in the wake of the 2003 invasion (an event that Hashem refers to as a 'liberation') and which, like the drawings, appropriate and address the war directly.[25]

Hashem's drawing practice originated with the 1991 war, when he began a daily series of sketches, which have since come to number some 2000 works on paper. While some of these drawings are held in the British Museum, most have never been exhibited. One work shown in the 2006 British Museum exhibition *Word Into Art*, was composed in Sweden at the outset of the allied bombing campaign, on 17 January 1991, and relates to this violent, punctual event as well as longer running, wider geopolitical developments. Hashem (2013) describes his approach to these drawings in the following way:

> They were made spontaneously, totally unconsciously, not planned. I made them while I was absorbed in watching and reading the news on the TV, internet, and other media. The symbols emerged unbidden from memories and previous experiences of conflict and life or death situations.

This particular drawing comprises depictions of several fantastical animals, including fierce birds that 'evoke both ancient mythical beasts and the American planes' (Porter 2006, p.111). As Hashem's note alongside the drawing records, there was a disjuncture between the experience of people outside Iraq, who had access to international media and people within the country who did not:

> I didn't sleep all night. I saw all the news reports, and I phoned all the Iraqis I know. None of them knew that the war had started… All the news so far has come from the West: America and the Coalition… There is nothing from Iraq about any military action … but the war will continue for more days… (Hashem cited in Porter 2006, p.111)

Hashem continued his series of drawings through the 2003 war. One such drawing (Figure 5.6), entitled *Buhriz*, deals again with both immediate, violent events and the circumstances of the war more broadly, and relates to telephone conversations the artist had one day with his father, still resident in Hashem's home city, as fighting was taking place between American forces and al Qaeda terrorists. As the Arabic text on the work relates, under the date 18 April 2004:

> I phoned my father today in Buhriz while the American bombing of the people… was going on for the second day. It was impossible for him to go out into the garden to get better reception on his mobile phone, and yet he took the risk and ventured out… God keep him safe. (Hashem in Ingram 2013)

As Hashem has further recounted of this event,

> He said, 'Look, I can't receive your call very well because there is fighting between the Americans and Al Qaeda. Our house is in the middle. So you can hear how the situation is. I have to go out to hear you, but I can't be there now, we are just lying on the floor.' … Ten hours later I called him, they had stopped fighting and he explained what was happening. (interview with author)

Figure 5.6 *Buhriz*, Satta Hashem (2004). Image courtesy of the artist.

The significance of the drawings that resulted from such events, and the manner in which they recall both a sense of co-presence with his father and memories of place, family and friendship for the artist, was underscored in the following way:

> This [experience] had a big effect on me and I did some drawings that day, very sketchy drawings. But these drawings for me are more important than anything, because it's like I'm sitting with him. I know this city, I know my village, I know this is our house, where we used to play. The destruction of a city is the destruction of my memory in fact. Everything is gone. I asked him about at least ten of my friends. He mentioned two only who are still alive. Most are dead. (interview with author)

This moving recollection adds to our understanding of the drawings not as depictions of something happening external to the artist, but as emerging from embodied practices that are interwoven with the event, both as part of the event and as a critical appropriation and memorialisation of it and of other events. Once again, we see that the relation of art to the event is multiple, differentiated and differentiating, not confined to a single moment, linear sequence or clearly defined topographic spatiality. What is evoked is a sense of entanglement as well as estrangement, and the continuing reality of the event.

This provides a basis to elaborate further on Caren Kaplan's (2013) formulation of the experience of the experience of war in Western countries. An important critical task in such contexts, I would suggest, is not just, as Kaplan argues, to recognise how war is felt and sensed by (implicitly, Western) people who are topographically and phenomenologically distant from war zones, but, first, to open up the sense of the 'we' to whom such propositions might apply, and second, to recognise the forceful, embodied ways in which many people with diasporic connections to countries in war experience it. This also moves us beyond Shapiro's (2012, p. 143) dichotomy between 'those who live the war zones' and 'those on the U.S. domestic front'. Artworks emerge from, and enact, immanent space-times that confound such dualisms.

Buhriz contains a number of motifs from Mesopotamian cultures and mythologies of the kind that surface throughout this chapter and the book, including the image of a winged bull amid twisted human and animal bodies, and the outlines of a cart wheel, evoking both the aesthetics of Assyrian civilisation and ancient Mesopotamia as a locus for the emergence of human technologies and societies, while also implicating the collections of institutions such as the Ashmolean and British Museums. As Hashem relates,

I mix the old and the new, superstition with the real, because if anybody told me about these Al Qaeda for example, slaughtering people, I would say, "Yes, but this is superstition, I can't believe it". ... The superstition has become real. And that is why I miss the old Mesopotamian art. They symbolise every abstract idea, they relate every abstract idea to animals, so wild animals for example are the devil and domestic animals are about peace. If you go to the British Museum and see the old Mesopotamian art you will see a lot of fighting between animals and from that mythology I took many of my ideas.

Hashem's drawing practice should be understood not simply as drawing upon ancient cultures to address contemporary events, however; it is rather that the art and artefacts of ancient Mesopotamian cultures have been intrinsic to the ways in which in which war, state formation and imperialism have been conducted, rationalised and legitimated in, and in relation to, Iraq, since their rediscovery in the nineteenth century.

In this sense, works like Hashem's *Buhriz* advance their own theorisations of the event. While Iraqi people can draw upon a unique, deep and rich set of cultural inheritances, then, and moreover in the context of their loss, destruction and abuse, the ways in which these are appropriated and channelled mobilise what Zainab Bahrani (2008, p.15) has called the 'magical technologies' of war. This phrase refers to those 'facets of war and domination that fall under the categories of representation and display, the ritualistic, the ideological, and the supernatural'

(Bahrani 2008, p.15).What Hashem's sketches and others like them further reveal, however, is how such objects and representations have served not only as resources for official propaganda, but ways of appropriating, working through and resisting geopolitical events and staking claim to one's own reality; that is, for their creative counter-actualisation.

The ways in which artworks operate as evental assemblages is also evident in relation to Hashem's painting practice, which explores diverse themes across portraiture, landscape, abstraction and geometric figures, informed by colour theory and ideas of the 'visual brain' (Hashem 2007a). As the artist has stated (Hashem 2007b), 'colour is not a physical thing; it is sensation, like feeling hot or cold'. Here Hashem's practice draws on his reading of scientific literature on the nature of mind, perception and aesthetics and in this regard his work signals the affective and more-than-representational dimensions to representation that have been explored by geographers following Deleuze's work on Bacon and Paul Klée (Dewsbury and Thrift 2005), further situating contemporary Iraqi art in relation to the transnational circuits of modernism. But while Hashem has explored the implications of such theories in many works not bearing any direct relation to the war or Iraqi politics, they have further resonances when it comes to those paintings that do address embodied violence, in which questions of visibility, exposure, seeing and recognition are intimately related to those of trauma, justice and politics. These works stage events in which such materialities, capacities and practices are in motion between work, viewer and world, drawing them into encounters with violent processes of becoming, but also ambiguous gestures of defiance, resistance and hope (Dewsbury and Thrift 2005).

Many of Hashem's works painted in relation to the 2003 war deal with themes of injury, death and mourning, including distinct series of works depicting grieving widows, exposed graves and body parts, while other paintings bear the title *Identity Unknown*. Hashem has related how he found the experience of watching the uncovering of mass graves in Iraq in 2003 and seeing people searching for their relatives deeply affecting.What I focus on here, however, is Hashem's 2007 work *Freedom No.1* (Figure 5.7), made as a part of a series that was commissioned and exhibited alongside the work of two other Iraqi artists (Thaer Ali and Anwar Badrie) in collaboration with National Museums Liverpool, Salford Museum and Art Gallery,Tyne and Wear Museums and Leicester City Arts and Museums Service. I conclude this part of the discussion with reference to the work in view of its evocation of the broader themes raised in the chapter overall.

As well as its partially overlapping title, this richly coloured and visually arresting painting bears certain formal similarities to Jawad Selim's Freedom Monument, in terms of its composition around recognisably human figures, and in conveying a sense of movement, struggle and tension.[26] But while this may reflect common reference points in Sumerian, Babylonian and Assyrian cultures, there are also important differences between the two works. *Freedom No.1* reads from left to right rather than right to left, but also moves in three acts, from the

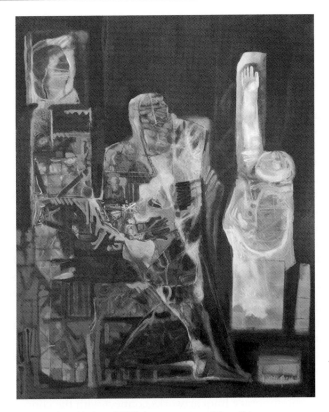

Figure 5.7   *Freedom No.1*, Satta Hashem (2007). Image courtesy of the artist.

confinement of a political prisoner on the left, to the struggle of a freedom fighter (rather than a soldier as in Selim's work) in the middle, to the birth of freedom on the right, in the figure of a child with an outstretched hand, while the use of the colour blue evokes its symbolisation in some cultures of hope. In clarifying the process through which this work was made, Hashem has further stressed that his training, resources and experiences outside of the country differ from those of artists who continued to live and work in Iraq. Like the Monument, however, this work can also be read as expressing a distinctly modernist aspiration in positing, against all events, that liberation might still be possible.

Conclusion

The works considered in this chapter bear out the argument by Nadje Al-Ali and Deborah Al-Najjar (2012, p.xxxi) that 'dislocation and displacement do not stop someone from identifying with, feeling for, and hurting about Iraq'. They also help to underscore the extent to which the geopolitical event of the 2003 war needs to be understood as multiple, in terms of how and where it was encountered, its relation to previous events and in terms of the different ways in which it has been appropriated and counter-actualised in artworks. Examining artworks and the trajectories of the artists who have created them in some detail and in relation to Iraq's art history as well as its political history also heightens an appreciation of the specificities of Iraqi experience and aesthetics. This further emphasises the ways in which Iraqi experience and aesthetics have become implicated in wider currents of modernism, but also contribute to them and call them into question.

Consideration of these works reveals multiple temporalities and spatialities enacted through art before, during and since the 2003 war, as artists have worked through feelings of estrangement and entanglement amid ongoing events. Conceptualising these workings-through as assemblages that enact counter-actualisations of the geopolitical event leads us to consider how they may both express the event and double it, detaching it from its self-evidence and effectively liberating it 'from the limits of individuals and persons' (Deleuze 2015, p.155). This offers the opportunity not just to contemplate a representation of an event, but in some sense, and however imperfectly, to immerse oneself within it and to go through it otherwise.

Considering these works collectively also points towards virtual events, or Events, beyond empirical experience, but which have been actualised and counter-actualised in diverse ways. This might include the war as an instance of the colonial present, but, drawing again on the work of Zainab Bahrani (2014), it can also be argued that, in working with elements of the country's ancient civilisations, many contemporary Iraqi artworks connect with the conceptions of deep time evident in reliefs and steles constructed thousands of years ago, thereby linking recent geopolitical events with the Event of human civilisation itself.

Notes

1   The analysis here in some ways echoes Doreen Massey's discussion of 'place beyond place' (Massey *et al.* 2009). I am grateful to the anonymous reader of the manuscript who highlighted this connection. There is scope to interpret things more fully in terms of relationalities than I have done here.

2   I am grateful to Nadje Al-Ali and Charles Tripp for conversations that shaped my understanding of the literatures and issues discussed in this chapter.

3   According to artist Satta Hashem, 'We had a kind of renaissance in the fifties and sixties where painting and drawing and writing all became a feature of the new Iraq. Big names appeared in Iraqi art and they gathered a lot of public attention' (interview with author). As he also recalls, 'After the Second World War and in the new political situation in the colonised countries, the self-awareness of Iraqis became very pronounced and they started to challenge the Europeans about their identity and their existence' (interview with author).

4   Tripp (2007) describes this event as a *coup d'état* rather than a revolution; as Gregory relates, 'Most Iraqis greeted the fall of the Hashemites with enthusiasm, though popular participation in the coup was largely gestural' (2004a, p. 151).

5   As well as a polemical critique of Baathist aesthetics (Makiya 1991), Makiya wrote two influential but widely critiqued books (Makiya 1989, 1993) arguing that Saddam Hussein's Iraq had become a totalitarian state whose victims had been abandoned by Western leftists and Arab intellectuals. One of the most forceful advocates of regime change in Iraq, he went on to play a leading role in the CIA-backed Iraqi National Congress run by Ahmed Chalabi (who is discussed briefly in Chapter 8), and returned to Iraq after the invasion in an attempt to shape the country's new politics. Having become disillusioned with the results of the invasion, Makiya returned to his university post in the US in 2006. Despite the flawed and contentious nature of much of his work, his discussion of the Freedom Monument (Makiya 1991) and its context accords with other discussions of the work and political developments around that time.

6   As well as organising the Rabat Biennale (Smith 2016), in this period Dia Al-Azzawi also participated in the First Triennale of 'World Art' in India and two editions of the Biennale Grafike in Ljubljana (Gardner and Green 2013). Azzawi left Iraq for London in 1976.

7   The poet Fares Haram provides an elegaic account of the politics of aesthetics in Iraq under the Baath and after 2003 (in Badr *et al.* 2015).

8   In this regard Al-Ali and Al-Najjar (2012) provide a sophisticated discussion of Iraqi identity and diaspora politics.

9   Malallah describes an environment under the Baath in which talented artists had no alternative but to exist under government control, and in which any underground or unapproved activity could be fatal: 'There was nothing outside'. While many family and friends left Iraq during the 1990s, she remained, though she had been opposed to Saddam, as she describes, 'from the beginning: there is no civil life, everything is military' (interview with author).

10 Malallah's post-2007 work has continued to address the themes of museums, heritage, memory and survival; the first place she visited on arriving in London was the British Museum (interview with author).

11 As Malallah describes, artists who emerged through Iraq's art institutions under the dictatorship of Saddam Hussein in the 1980s were effectively 'trapped' in the country as they could not travel and had to endure three wars: Iran-Iraq (1984–1988), the 1990–1991 war, and the 2003 war (Asfahani 2016). For her, the pivotal event in the modern history of Iraq was Saddam's invasion of Kuwait, which set in train a sequence of events leading to the collapse of the country after 2003.

12 Artist talk at Chelsea School of Art, 23 May 2012.

13 Some of these works have subsequently been purchased by private collectors. Two of Malallah's artist books (a form often employed by artists from Iraq, and especially by artists within Iraq under sanctions) are held in the British Museum.

14 *My Country Map* has also been denoted by the slightly, but significantly, different title *New Map/US Map* (Al-Baholy, in Malallah 2009).

15 For the artist, *My Night* refers not only to the initial bombing but also to the 13 February 1991 bombing of a shelter in the Amiriyah district of Baghdad in which hundreds of civilians, mostly women and children, had taken refuge. Two 2000-pound bombs were dropped by F117A aircraft on what US officials subsequently

claimed they had believed was a 'leadership target', but which they had previously acknowledged had been used as a civilian refuge during the Iran-Iraq war. While reports in the following weeks and months stated that 200 or 300 people were killed in this event, later reports put the figure at 408 (Human Rights Watch 1991; Arango 2016). Artist talk, Chelsea School of Art, 23 May 2012.

16 Al-Attar associates views of the River Thames, close to which she has a studio, with the Tigris in Baghdad: 'I long for the sky, I paint the sky, I look at the River Thames and say, this is Baghdad, the Tigris. It's always the sunset' (interview with author).

17 Al-Attar has subsequently returned to the colours and themes of her earlier work.

18 Al-Attar lost both her parents, who had remained in Baghdad, in wartime; her father in 1991 and her mother in 2004 (interview with author).

19 This is also a recurring focus in the work of Dia Al-Azzawi (also discussed in Chapter 4).

20 This work was exhibited at the Leighton House gallery in the Holland Park area of London in 2006; images of the work appear in a catalogue for this exhibition (Leighton House Museum 2006).

21 The collective project was exhibited at the Charnwood Museum in Leicester in March 2004 (BBC 2004). An undated exhibition catalogue (*Black Rain: Reflections of War*) explains how participants experienced the war: 'They lived this suffering through their families inside Iraq still, or by themselves as exiles far away from the country from which they had escaped.'

22 Naser further relates that when the *Black Rain* project was exhibited in Damascus, the title aroused criticism from pro-invasion Iraqis who argued for simply *Rain* (interview with author). Naser's account recalls the manner in which great floods such as that described in the *Epic of Gilgamesh* have often been taken as points defining the beginning of the 'present'; important artefacts that record the *Epic* dating from Babylonian and Neo-Assyrian times are held in the Ashmolean and British Museums.

23 Naser had to leave the Lebanon in 1991 when the PLO cancelled his visa. While the PLO supported Saddam's invasion of Kuwait, Iraqi exiles in Lebanon did not.

24 As Deleuze and Guattari (1994, p.110) write,

> books of philosophy and works of art also contain their sum of unimaginable sufferings that forewarn of the advent of a people. They have resistance in common-their resistance to death, to servitude, to the intolerable, to shame, and to the present.

25 As Hashem has related,

> What happened is that what should have been a liberation, became a nightmare for everybody. In fact possibly 90 per cent of Iraqis who live inside or outside of Iraq, we were very optimistic, before the war. We supported the war, because it will end the tragedy of the Iraqi people. But later something happened that nobody could control, and yesterday's victims became the new criminals. So it's like nothing happened – you see the same film, with different actors. (interview with author)

26 Here I am indebted to clarifications and additional detail provided by the artist by email. I would emphasise that the Freedom Monument is neither reference point nor inspiration for *Freedom No.1;* the decision to link the discussion of the two works in this way is mine alone.

Chapter Six
Geopolitical Aesthetics of Oil

British and American politicians and officials strenuously denied that oil had anything to do with the decision to invade Iraq in 2003. Shortly before the invasion, Prime Minister Tony Blair stated in a live television interview that 'the oil conspiracy theory is honestly one of the most absurd when you analyse it' (BBC News 2003b). Such denials echoed the protestations of British politicians involved in carving up the Ottoman Empire in the wake of the First World War. Winston Churchill stated in the House of Commons that the idea that the military campaign by which British control in Iraq had been established had been motivated by oil was 'too absurd for acceptance' (cited in Muttitt 2011, p.4). In a speech given to the Lausanne Conference of 1923 concerning the British argument that the former Ottoman province of Mosul be included in the new polity, Lord Curzon stated:

> It is supposed and alleged that the attitude of the British Government to the *wilaya* of Mosul is affected by the question of oil. The question of the oil of the Mosul *wilaya* has nothing to do with my argument. I have presented the case on its own merits and quite independently of any natural resources that may be in the country. (cited in Sluglett 2007, p.72)

It would have been strange indeed had oil not been a factor in either case. By 1923, interests within the British Navy had been pushing for more than a decade to convert the fleet from coal to oil, an initiative that Churchill himself had previously championed as First Sea Lord. A corollary of this policy was held

Geopolitics and the Event: Rethinking Britain's Iraq War Through Art, First Edition. Alan Ingram.
© 2019 Royal Geographical Society (with the Institute of British Geographers).
Published 2019 by John Wiley & Sons Ltd.

to be that Britain had to own or control a significant proportion of the world's oil resources. Shifting the strategic focus from coal to oil would, it was envisaged, also reduce dependence on coal workers at home, whose growing influence also troubled Churchill and other leading politicians (Mitchell 2013). Though Blair sought to deny that oil was a motivating factor in the move towards the 2003 invasion, regime change in Iraq and enhanced American investment and involvement in oil-producing countries had formed two key policy priorities for the incoming George W. Bush administration, and in the run up to the war British oil companies quietly moved to position themselves as advantageously as possible in anticipation of a post-Saddam Iraq (Muttitt 2011).

While much political analysis has addressed the machinations and manoeuvering surrounding the 2003 invasion, writers in geography and related fields have offered several more theoretically-oriented critiques of the war and the geopolitics of oil surrounding it, with interpretations focusing on the geopolitical economy of US-centred imperialism (Harvey 2003), military neoliberalism (Boal *et al.* 2005), the resurgence of Orientalist imaginaries and sovereign violence after the 9/11 attacks (Gregory 2004a) and the material politics of sociotechnical networks amidst concerns about peak oil (Mitchell 2013). Rather than offering another explanation, what I seek to draw out in this chapter are some of the ways in which oil has been appropriated and reworked aesthetically in relation to the event of the war. A consideration of the appropriation of oil in artworks emerging from the 2003 war, I suggest, draws us further into a genealogical way of thinking about the war, while enhancing a sense that this was a war that was at least in part not just for oil, but *of* oil, thereby also positioning it critically as an anthropocenic event (Bonneuil and Fressoz 2017). The ways in which oil and other materials are used in these artworks also call the human body into question in a manner that, while highly attenuated and localised, evokes the embodied violence of geopolitical events.

Oil has been implicated in the geopolitics of Iraq in numerous ways. In 2003, control over Iraq's energy resources and infrastructure was a central war aim for the US and its allies, but the invasion and occupation themselves also consumed vast quantities of petroleum, and oil, fire and smoke spilled, erupted and drifted across the Iraqi landscape. With the US and Britain intensely concerned about oil wells being set on fire and crude oil being dumped into bodies of water, as had taken place in 1991, Iraqi forces also sought to use oil for defensive purposes. On the third day of the coalition attack, the Baghdad-based blogger Salam Pax wrote:

Half an hour ago the oil-filled trenches were put on fire. First, watching al-Jazeera they said that these were the places that got hit by bombs from an air raid a few minutes earlier, but when I went up to the roof to take a look I saw that there were too many of them – we heard only three explosions. I took pictures of the nearest. My cousin came and told me he saw police cars standing by one and setting it on fire. Now you can see the columns of smoke all over the city. (Saturday 22 March 2003 in Pax 2003, pp.129–130)

In the wake of such actions, and in resonance with events during the 1991 war, burning oilfields and thick plumes of smoke quickly became one of the key visual motifs of media reporting. As I explore in this chapter and elsewhere in the book, many artistic and curatorial projects have addressed the geopolitical aesthetics of oil more critically, enabling people to encounter its implication in the ongoing event in many different ways. Artists who incorporated oil directly into works include kennardphillipps (for example in their 2004 *Award* series, made using oil, dust and blood), Hanaa Malallah (in *My Country Map*, 2008 and *Happened in the Dawn*, 2011, mixed-media works using layers of burnt canvas and oil), Andrei Molodkin (*Liquid Modernity*, 2009, a sculpture composed of a cage-like structure of acrylic pipes, through which crude oil is pumped) and Emily Johns (in the *Conscious Oil* project, which juxtaposed linocuts made with ink and oil alongside poetry by Anne Rouse and writing by Milan Rai). In the 2014 exhibition *Crudification* at P21 gallery in London, nine Iraqi and two British artists worked with oil barrels to create mixed-media sculptures and video works. The broader ecological and geopolitical ramifications of oil have continued to be raised in art-activist interventions by the Platform group (Barry 2013), and from 2010 the Liberate Tate group opposed oil company BP's sponsorship of cultural institutions by staging a series of unauthorised performative works in and around Tate Britain and Tate Modern (Miller 2015; Ingram 2017). Each of these works, in different ways, can be read not just as counter-actualisations of the war but as addressing the intersections between coloniality, militarism and oil in the anthropocene.

The chapter proceeds in the following way. I first consider Rashad Selim's mixed-media assemblage *Souvenir from the Ministry of Justice*, which incorporated a cartographic representation of Iraq's oil reserves made from recovered plastic, and which formed part of a series of works created in The Hague that addressed the question of international law as well as the role of oil in the war. I then broaden out the parameters of the event, considering how the growing role of Britain in Iraq from the late nineteenth century onwards was shaped not just by the country's strategic location with regard to India – a key initial motivation – but increasingly by a sense of the likely future importance of oil. Here I develop an account both of how oil figured in the emergence of Iraq as an experiment in late colonial government, and the ways in which the region was represented in artworks created by British military personnel more than 80 years before the 2003 war. In light of this discussion, I then consider two further projects that have explored the aesthetics of oil since 2003. The first of these is Richard Wilson's oil-based sculptural installation *20:50*, versions of which were from 1991 to 2016 continuously exhibited in galleries owned by Charles Saatchi in London, but which in 2009 was also realised in a former Baathist security prison in Sulaymaniyah in Iraqi Kurdistan. I then consider *Black Smoke Rising*, an immersive sculptural installation by British artist Tim Shaw that, like *20:50*, is organised around a large pool of oil, while referring more directly to embodied geopolitical violence and trauma. In enacting oil differently from the ways in which it is typically experienced in Britain, each work presents distinct ways of appropriating, reassembling and encountering the war as geopolitical event.

Souvenir from the Ministry of Justice

In late 2007, Rashad Selim was working as artist in residence the Gemak gallery in The Hague as part of a programme examining the splintering of urban space under conditions of war and security, much of which dealt with ongoing events in Iraq and Afghanistan and the war on terror more generally (Kluijver 2010). Usually resident in London, Selim had also lived and trained as an artist in Iraq, and in the 1990s became involved in the International Network for Contemporary Iraqi Artists. While much of his work since 2003 was initially concerned to oppose the invasion and occupation, often engaging with the problem of oil, Selim has shifted his focus to the development of projects involving Iraqi craft workers and materials that seek to revive traditional cultural and technological forms to address the ecological as well as political situation in Iraq and beyond.[1] Here I focus on the project Selim pursued at Gemak and one work in particular, *Souvenir from the Ministry of Justice* (Figure 6.1).

Souvenir... was an installation composed from materials mostly recovered from a demolished London supermarket. What might initially have appeared to be a chaotic collection of random objects can be seen as enacting polemical opposition to the war, while offering a genealogical framing of its causes, conduct and consequences. *Souvenir...* can be seen as an event that intersected with the event of the war, while addressing specific events within it and situating itself in relation to longer-term occurrences. Extending across the gallery and containing references to other works by the artist within the exhibition and the city, the work drew the visitor in via a series of visual cues, while also directing them outwards to a consideration of The Hague as a centre of international law and the oil industry.

The centrepiece of the work was a partially collapsed set of shelving. One more-or-less surviving shelf was strewn with black and white plastic cable ties, while upon the shelf below rested a bundle of reeds wrapped and tied in a black shroud, surrounded by fragments of wood. Two pairs of metal pipes also rested precariously behind the shelving, while a series of elongated black plastic patches ran in a loose, linear arrangement across the gallery floor, through the shelving and up the wall.

Each element of the work bore some connection to the war. The cable ties were of the sort commonly used in construction and DIY projects, but which had been appropriated by US and British forces for their usefulness in securing the limbs of detained people. These mundane objects can be seen as a reference to counter-insurgency techniques, while the collapsed shelving and wood readily evoked the destroyed furniture widely seen in images of bombed and looted buildings, another concern of the artist (interview with author).

Figure 6.1   *Souvenir from the Ministry of Justice*, Rashad Selim (2007). Photo by author, courtesy of the artist.

The title and form of the work drew attention both to what the artist saw as a lack of justice and due process for citizens of Iraq, and to a particularly contentious event within the invasion, when, upon assuming control of Baghdad, US military forces had provided protection only for the Ministry of Oil, while other government and public institutions were attacked and looted. *Souvenir…* thus made a direct, polemical statement about the injustice of the war and the conduct of the invading forces. The black patches, meanwhile, replicated patterns seen in geological maps of Iraq's oil reserves that had been widely circulated in the media in advance of the invasion. As the artist later explained,

Alan Ingram: You addressed destruction in the Ministry of Justice installation with this theme of destroyed institutions as well as the destruction of heritage, documents and the use of documentation…

Rashad Selim: Documents, yes, but also what I was seeing as the inversion of geography if you will, [where] what is underneath, what is hidden, becomes what's above. What's above is buried underneath…, that's what happening. *[Souvenir…]* had the map of Iraqi oilfields [across] the space. It appeared more and more, to me and to other people, that, really they didn't have any care whatsoever for what may happen, for the country itself. That includes its past as well, its ancient past as well as its contemporary past. Nothing… there was total disregard for everything. There was no regard for anything. So to express that, and to say that this *is* what is happening – there's no regard for our gene pool, there's no regard for our water, there's no regard for the agriculture, no regard – a disregard, a dissing of everything, including those laws and conventions that unite us in the spirit of modernity. (interview with author)

What became clearer on closer inspection was that the black shapes were accompanied by small captions displaying the names of Iraqi cities, providing a topographic connection to Iraq's geography beyond the more abstract topology of the blobs, while the black piping arranged behind the shelving provided a further tie between the black marks and the rest of the installation.

While cartographic representations of supposed oil deposits within Iraq and elsewhere are often conjoined with future-oriented invocations of economic potential, *Souvenir…* detached them from a technical context and implicated them in failures of justice and the absence of law and order. Selim also raised this issue in a series of public space interventions around The Hague that highlighted its role as a centre of international law and the global oil industry (Ingram 2009). Selim stated that he aimed 'for an essential naiveté through works directly involved in political and legal acts' (cited in Kluijver 2010 unpaginated), acts that included protests and other interventions around the city.

While the representation of oil reserves via black plastic shapes provided one connection to these interventions, another link was provided by the bundle of reeds. As Selim (2018) has explained, the reeds allude to the emergence of agriculture and settled civilisations in the region several thousand years ago and embody intimate ties with its landscapes, environments and cultures. Bundles of reeds were used to construct the round coracles (or *guffa*) that served for centuries as a principal means of river transportation. When sliced through, a bundle of seven reeds would also produce a hexagonal pattern of six circles arranged around one in the centre, linked by Selim to the auspicious, non-denominational 'seven eyes' symbol. Selim reproduced this symbol on a moveable road sign, which he then, in a further call for justice, deployed at important sites around The Hague. The artist also created a graffiti stencil of the symbol, and one evening sprayed it around the city until he was stopped by the police, whereupon he again called for enforcement of international law and the payment of compensation for damage done to Iraq (see Ingram 2009).

Souvenir for the Ministry of Justice thus provided a direct response to the lack of observance of law and the destruction caused by the invasion and occupation, but also offered a model for exploring them as complex geopolitical events. When viewed through its connections to the more performative works created by Selim around this time, the work can be seen not just as a critique, but a counter-actualisation of the war, in that it used the forms and practices of art to make something else of the event, and, in moving from the gallery to the street and back again, blurring the boundaries between art, politics and law.

Oil, Modernity/Coloniality and the Formation of Iraq

The first recorded journeys of British people in Iraq were made in Elizabethan times by diplomats and spies.[2] After a series of missions by British agents, nine or ten representatives of the Levant Company, which had been created to exploit trading opportunities in the region, passed through Tripoli, Aleppo, Bir, Baghdad, Basra and Hormuz on their way to India, one merchant returning via Mosul, Mardin and Orfa as well as Baghdad and Bir (Saleh 1966). In 1763, the East India Company established a 'factory' (or trading location) at Basra (Morton 2006: 8); trade expanded further after the construction of the Suez Canal in 1869, and the development of steamships during the nineteenth century facilitated greater access to southern Iraq via the Tigris and Euphrates (Sluglett 2007). As European tourism increased to Egypt, Palestine and Constantinople, however, Mesopotamia initially remained the preserve of the more adventurous.

Growing British involvement in the region was motivated by several factors. One was the prospect of further trading opportunities, but Mesopotamia came to be recognised as being of great strategic significance in view of its location between Europe and India. Archaeologists, geographers and explorers also came to be motivated by a belief that the region contained Biblical sites, including the Garden of Eden, and, as Priya Satia (2008) argues, by a search for aesthetic experiences that might relieve their disenchantment with modern life.

A dawning sense that the region might contain extensive petroleum reserves gradually transformed its significance in the British geopolitical imagination. A study by a Captain F.R. Maunsell, published in *The Geographical Journal* in 1897, concluded with the following vision of its promise:

> Possibly one result of the present political troubles of Turkey may be a greater facility of obtaining concessions to develop some of these remarkable mineral riches on modern lines, and if properly explored, there is no doubt that the Mesopotamian petroleum field might be made to yield results of the greatest importance. (Maunsell 1897, p.532)[3]

The growing importance of petroleum meant that Mesopotamia came to occupy an increasingly significant position in relation to what Satia (2008, p.19) calls the 'spy-space' of Arabia. At the outset of the First World War, an expeditionary force of 5500 troops drawn mostly from India was deployed under imperial command to Basra to protect British interests. The force and its mission were quickly expanded, however, and, despite setbacks, by 1918 Britain had occupied all three of the Ottoman *vilayets* of Mosul, Baghdad and Basra, establishing administration in the occupied provinces based on the model developed in India (Dodge 2003). Britain then assumed post-war control of these territories under the Mandate system, and, in the face of regional pressure and uprisings among the Shia populations along the Tigris river, sought to bring them together in a single polity.

With international affairs increasingly shaped by the Wilson administration's efforts to develop a liberal system around ideas of self-determination and economic openness, neither continued occupation nor direct rule were viable options for securing British interests (Sluglett 2007). Facing severe financial pressures, Britain also had to compromise with other European imperial powers (principally France) and with the new Turkish state emerging from the collapse of the Ottoman Empire. In this context, British officials embarked upon what Leo Amery, Secretary of State for the Colonies, called a 'novel experiment' (cited in Sluglett 2007, p.64), as the new state of Iraq came to be envisaged in turn as a model for late colonial government. Amidst internal debates about whether the region should be managed from London, Cairo or Delhi, officials promoted the idea that British trusteeship would provide a vehicle for Iraq to progress towards independence, while seeking to secure British interests and maintain Iraqi dependence by installing advisors in key government ministries and using RAF air power to intimidate local populations.[4] British control also rested on an alignment of interests with predominantly urban, Sunni elites who were relied upon to form the core of the government. Both parties considered the inclusion of the Mosul *vilayet* in the new state to be essential in order to ensure Sunni pre-eminence in government, thereby ensuring British access to its oil reserves.

The matrix of modernity/coloniality into which the region was drawn and through which the modern state of Iraq emerged had aesthetic as well as political-strategic dimensions. While the region was being explored, surveyed and mapped, it was also being sketched and painted. Though British Orientalist painters of the nineteenth century had tended to find their inspiration in Egypt, Turkey and Palestine, some more adventurous artists travelled through Mesopotamia. They included the Scottish painter Arthur Melville, who spent six weeks in Baghdad (National Galleries Scotland 2017a) before travelling on to Constantinople, a journey on which he was robbed, left for dead, and imprisoned as a suspected spy. Melville's time in Baghdad provided a basis for romanticised watercolour depictions of everyday urban life, a form of representation largely replicated by artists stationed in Iraq during the Second World War and by some who were embedded with British armed forces in Iraq in 2003 and after. The British armed forces that moved to assert control in the region during and after the First World War also included artists. Sydney Carline, who had trained at the Slade and served as an aviator in Europe during the war, was in 1919 commissioned by the new Imperial War Museum (IWM) to record aerial warfare in the Middle East. In this commission, he created a series of sketches, studies and watercolour and oil paintings recording the British campaign against Turkish forces, including paintings of aerial attacks in Palestine and Jordan, and scenes from Baghdad and from northern Iraq during a brief period stationed in Mosul.

One of Carline's best known works vividly illustrates the colonial military technologies and techniques that would later become central to British efforts to control the new state. *Over the Hills of Kurdistan: Flying Above Kirkuk* (1919), an oil version of which is held by the Ashmolean Museum at Oxford University and a watercolour by IWM London, show Royal Flying Corps biplanes returning from a bombing raid, with a view from one plane backwards over the town of Kirkuk and surrounding fields and roads, against a backdrop of hills and mountains. These works thereby register both the extension of British power into former Ottoman territory and a colonial way of seeing from above (Adey *et al.* 2013). There is a vertiginous as well as panoramic quality to the view from the painter's plane, which is angled to one side, and the way in which the other planes are pitched above the land expresses a circling motion that might be playful as much as tactical. What is missing is the violence such planes were used to enact; having watched a demonstration of bombing conducted on an 'imaginary village', Gertrude Bell wrote, 'I was tremendously impressed. It's an amazingly relentless and terrible thing war from the air...' (Bell 1927, letter of 2 July 1924; also Gregory 2004a, p.149).

Over the Hills of Kurdistan… depicts a form of military technology reliant upon oil-based fuel and thus captures something of the shift towards a petrocentric geopolitics. Here it is possible to make a connection with a near-contemporaneous oil painting, *The Navy in Baghdad* by Donald Maxwell (1918), which provides a view of the harbour at Baghdad.[5] In the foreground are a wooden sailing boat and three large coracles of the kind that are historically distinctive to the Tigris and Euphrates (and similar to those which Selim has constructed and used). In the middle ground, in front of domed buildings, palm trees and a minaret, are two wooden Royal Navy boats; behind them are two larger ships with funnels, out of which belch plumes of dark smoke. This work also, then, depicts the intrusion of military vehicles fuelled by carbon-based sources well into Mesopotamia.[6] In another Carline painting, *A British Pilot in a BE2c Approaching Hit Along the Course of the River Euphrates* (1919) (held, again, by IWM London) a thick plume of smoke billowing up from the ground is visible, from a position behind the pilot, who is looking to across the left. Also visible are green fields adjacent to the river, which winds its way from the foreground into the distance.

20:50 in Britain and Iraq

Oil's presence in and influence over the landscapes of Iraq is both material and mythical. Among the petroleum deposits represented in Selim's *Souvenir From the Ministry of Justice* is a 'supergiant' oil and gas field that runs close to the northern Iraqi city of Kirkuk, and which seeps through to the surface in many places. A short distance outside the city is a site known as Baba Gurgur, a rough hollow of a few square metres where gas reaches the surface and ignites, producing an 'eternal flame' that has reputedly burnt for thousands of years. Baba Gurgur has been taken to be the 'fiery furnace' described in the Book of Daniel, into which King Nebuchadnezzar cast three men who refused to bow before him, and the source of oil described by Plutarch as having been used for street lighting 'in order to impress Alexander the Great' (Morton 2006, p.8).

Exploratory drilling began in earnest in Iraq in the 1920s, and in late October 1927, a test well near Baba Gurgur drilled by the Turkish Petroleum Company (the forerunner of the Iraq Petroleum Company created in 1928) erupted in a gushing fountain of oil. According to one eyewitness to the containment operation, '[f]or hundreds of metres around everything was smothered in oil… The oil fell evenly in clouds all around the derrick and draw-works due to the windless autumn days' (unknown author, cited in Morton 2006, 10). In the following weeks, thousands of barrels worth of oil flowed across the landscape, and the Company had to improvise a series of dams and pools in an attempt to control the disaster. The risk of fire was constant and pockets of blue gas formed in the shallow depressions of the landscape, on one occasion killing three sleeping workers.

If oil has long been a pervasive presence in the landscapes surrounding Kirkuk, for residents of Britain and other Western countries the substance is for the most part contained and channelled by infrastructures and technologies intended to keep it as far as possible out of human contact; what people encounter most of the time are the products of its combustion and transformation. From this perspective, multiple instances of Richard Wilson's sculptural installation *20:50*, which takes its title from the name for a specific fraction of oil that was widely used in cars in the mid-to-late twentieth century, have offered an unusual opportunity to encounter oil directly. First created at Matt's Gallery in London in 1987, *20:50* was purchased by the Saatchi Galleries in 1991 and was continuously exhibited at its galleries until its sale to an Australian collector in 2016. In 2003, the work was installed at the Saatchi Gallery in County Hall on the South Bank of the River Thames in London. The work was then installed at the new Saatchi site in Chelsea in 2010 until its sale.[7]

Wilson's practice typically involves using engineering and architectural techniques to alter buildings in ways that challenge perceptions of interior and exterior spaces and of the materials from which they are composed. *20:50* takes the form of a single room filled to waist height with recycled sump oil. In the work, the oil is contained by riveted steel panelling; these materials are also used to form a walkway cut into the middle of the pool so that viewers can experience feeling surrounded by the oil. The installation offers a multisensory experience: the smell of the oil pervades the room, while its surface is reflective, reconfiguring the distribution of light and one's sense of orientation in the space. Depending on the dimensions of the room and one's standing position and visual attention, the space can appear deeper than the floor, contributing to what, along with an apprehension of immense mass and viscosity, can be a disorienting experience. The room appears much bigger than it ought to, while the apparent volume of oil exerts an ominous sense of barely contained force. After a time, the smell of oil becomes unpleasant and induces a need to leave the space. Engaging olfactory, visual and haptic perception, human embodiment and consciousness are set in motion and thrown into question by the work.

20:50 might be read in a polemical way as confronting viewers with a substance that enables their unsustainable way of life, while destroying environments and fuelling conflicts around the world. However, such associations and contexts have not been highlighted in the presentation of the work, by the Saatchi Gallery or in existing critical appraisals (Morrissey 2005). As Wilson has also stated (in Morrissey 2005, p.18), oil was chosen primarily for its reflective qualities. In Central London, *20:50* could function as an experience of materially mediated embodiment, perception and contemplation, without attention necessarily being called to geopolitical problems. Indeed, for Morrissey, the work in some ways embodies the promise of autonomous art: 'Wilson does not ask us to imagine: instead he makes concrete interventions in the physical world in order *to make a different order truly palpable*' (2005, p.7, emphasis added). But the work is not understood as being fully autonomous; in the words of the artist himself, 'I have to go to a real thing... I need that initial thing *from the real world* because I've always been concerned with the way you can alter someone's perception, knock their view of the world off-kilter' (cited in Morrissey 2005, p.7, emphasis added). *20:50* has further been understood as being aligned with the strategies of conceptual art, in that it is 'essentially an idea' (Wilson in Morrissey 2005, p.18), a 'set of conditions applied to a given space rather than a work that is *about* that space... [it] is essentially experiential' (Morrissey 2005, pp.18–19). In creating a new material and perceptual environment, the work is disruptive of the relation between body, mind and world, of sensation and sense making, and it can in this sense itself be regarded as an event, one that is ostensibly autonomous but irrevocably bound up with the conditions of its existence and thus, ultimately, the anthropocene.

The work was also realised in 2009 in the city of Sulaymaniyah, a hundred kilometres from Kirkuk in Iraqi Kurdistan, where it developed a series of new, more explicitly political resonances related to site and context. The form of the work here closely replicated the previous instances in London: a single room with one wall taken up by pillars and exterior windows was filled by oil, secured to waist height by riveted steel panels, again with a walkway cut into the middle, enabling individual viewers to advance to the centre of the installation. A photograph (Figure 6.2) conveys something of the deceptive visuality of the work, as the windows and pillars at the far side of the room appear to extend below the level of the camera, creating the impression of a much deeper volume. The 'floor' (a reflection of the room's ceiling) appears to be several feet below the surface of the walkway extending into the middle of the oil pool and the walkway seems more like a balcony extending into space over the floor beneath. When viewed from a different angle, the perspectival effect of floor and side panels converging towards a rectangular panel at the far end looks more like a passage leading down to a door well below eye level.

Figure 6.2 *20:50* installed at the Red Jail, Richard Wilson (2009). Photo by Mark Terry, courtesy of the photographer and artist.

A repetition of an existing work, installed in a room of similar dimensions and configuration, ostensibly offers the same kind of embodied encounter with materials and with space. Indeed, ArtRole, the organisation that arranged the installation described the work in a way that implied its status as an 'immutable mobile' (Latour 1987): '[a]n internationally acclaimed installation by renowned British sculptor Richard Wilson RA was shown for the first time in the Middle East'.[8] At the same time, the organisation and Wilson himself also recognised new geopolitical dimensions to the work.[9]

20:50 was exhibited as part of the Post War Festival, the title of which can in this context be taken as referring both to the 2003 war and the war that Saddam Hussein had waged against Iraqi Kurdistan until the 1990s. The festival's name thus constructed these wars as events that had passed. In 2003, regional authorities had moved to enhance the autonomy they had gained as a result of the 1991 war. For several years, areas controlled by Kurdish forces had by and large avoided the kind of wide-scale armed violence taking place elsewhere. The festival was supported by the British and US Embassies in Iraq, the British Institute for Iraq, and a local oil company and media group (ArtRole 2009).[10] The festival was centred in the Amna Soraka, or Red Jail, a complex of buildings that had been constructed under Saddam Hussein as part of the campaign to repress Iraq's Kurdish population and strengthen Baghdad's control over the region. Hundreds of people were tortured and executed at the site between 1979 until 1991, when Sulaymaniyah was liberated by Kurdish forces, and in 2000 the site became a museum of war crimes.

Here *20:50* was both separated from and entangled with its surroundings and contexts in a way that is highly suggestive. On the one hand, as an installation within a defined and highly controlled architectural space, the work is again constructed as a space of encounter and contemplation, where a body can approach and interact with a material in ways that are quite different from those typically on offer in everyday life. The work was thus again constituted in great measure by its separateness from life and from politics. That the work could again come into existence in this way testifies to the operation of the aesthetic *dispositif* of art in constituting a distinct sensorium. For the work to be the 'same' work in both London and Sulaymaniyah – for it to function as an immutable mobile – therefore requires some measure of disavowal of setting and of events. At the same time, having hitherto existed mostly within the contemporary art world in Britain, the work could not but be transformed by its subsequent installation in a place that, while it should not be reduced to its place in the geopolitics of oil, has long been on the frontlines of what Mitchell (2013) calls carbon democracy. Amidst the geopolitical event of the 2003 war, the simultaneous existence of *20:50* in the Saatchi Gallery in London and the Red Jail in Sulaymaniyah established a temporary topological connection between the two places, linking them in hitherto unanticipated ways and gesturing towards their virtual entanglement.

Black Smoke Rising

The sculptural installation *Black Smoke Rising* by Tim Shaw also works with and through oil.[11] This work is remarkable among responses to the war in Britain, I suggest, in terms of how it mobilised sculptural and installation practices to create an intense, immersive and deeply troubling experience of geopolitical power and violence. The work contained numerous material and symbolic references to the Iraq war as event, but also implicates it and the person experiencing the installation in a much wider set of events, environments, practices and relations. Once again, I argue that the work can be interpreted as a counter-actualisation of the war in that it offered both a model of the event and a way of encountering it markedly different from that available in the worlds of politics and the media. I will again describe how the work emerged out of events, while itself being an event and giving rise to further events as it was encountered.

As well as offering this conceptual framing, some critical contextualisation is necessary in order to situate *Black Smoke Rising* in relation to what I have conceptualised as the distribution of sensibility. The work raises issues in terms of the appropriation of images of violence, and how conflicts whose violence is concentrated 'elsewhere' are 'brought home' to 'domestic' publics, whose impressions of the event come primarily through media reporting. In also evoking the complicated political geography of Northern Ireland and the United Kingdom, however, the work also questions easy distinctions between 'here' and 'there'.

If one of the prevailing refrains in accounts of war and artistic responses to it is that it is impossible to know what war is like without experiencing it directly, *Black Smoke Rising* sought to stage something akin to this experience. Here, again, it is essential to consider the material and affective qualities of the work. In particular, the formal properties of installation art are important in that, in providing an immersive environment, installations challenge ideas of what the spectator is and does, and of what they experience. Installation can be haptic, auditory and olfactory as well as visual, and, as Harriet Hawkins (2010) has explored, is spatially and temporally encompassing of the body of the viewer/participant. Installation provides an experience that is multisensory, which is entered, encountered, and exited, but which may endure in bodily sensations as well as cognitively appropriated memories. While the large scale installation forms that have proliferated in the international art world since the 1990s have sometimes been criticised (for example Bishop 2010) as providing spectacular entertainment, a notice and disclaimer at the entrance to *Black Smoke Rising* warned visitors that they might find it not just 'affecting' but 'disorienting', that the oil and charcoal used presented risks to clothing and health, and that visitors might find the smaller installation within it, *Soul Snatcher Possession*, 'disturbing and claustrophobic'; visitors under 16 had to be accompanied by an adult.[12]

The environment that *Black Smoke Rising* created was not only unpleasant and confronting in terms of its materiality, however; it also contained works evoking images of embodied violence and suffering, and two widely-circulated photographs from the war in particular. The first was the notorious picture taken of a hooded Iraqi man standing on a box at Abu Ghraib prison in 2003, which came into the public domain in spring 2004, which was rendered sculpturally as *Casting a Dark Democracy*. The second was a photograph of a British soldier, in flames, in the process of escaping from a tank at which a petrol bomb had been thrown during a riot in Basra, a photo taken by a press photographer and widely reported in 2005, rendered as *Man on Fire*. One of the stakes of each sculpture and of the installation overall, therefore, was that evoking and confronting these images, and exploring their broader historical and political resonances (Khanna and Shaw 2015), was aesthetically and critically necessary in terms of bringing war more affectingly into the public domain in Britain.

Black Smoke Rising was comprised of the two massive sculptures positioned within within an immersive environment, which was organised as a whole around a large, black, reflective pool of motor oil, with a further sculptural installation presented in a small, oppressive room in one corner of the space.[13] Here I focus on the main space, in which oil figures most centrally and which bears the clearest relation to the Iraq war, and in describing briefly my own experience of the work, switch temporarily into the present tense.

Figure 6.3 *Black Smoke Rising,* Tim Shaw, MAC Birmingham (2014). Image courtesy of the artist.

The experience of the installation begins before it is encountered directly: a heavy, rhythmic clanking sound, and the smell of oil fumes are evident outside the exhibition space (Figure 6.3). Upon entering, the volume of sound and the intensity of the smell of oil increase markedly. The sound is like a mix of a bass drum, a heartbeat and harsh metallic noise, repetitive and jarring. The sense of disorientation is compounded by the dimness of the space and a thick mist, which impede vision and contribute to an unpleasant sense of heaviness, oppression and foreboding. Black hand marks are daubed around the walls; walking around the edge of the space, some of them became legible as words: IRAQ, DEMOS, KRATOS. The smell comes from a large black pool of oil in the middle of the room, surrounded by sand, taking up most of the floor area and leaving only a walkway around the edges. The pool reflects its surroundings, including a gigantic figure standing its far end, a figure that immediately recalls the hooded 'man on a box' in photographs of the torture at the Abu Ghraib prison complex (Figure 6.4). Positioned off on the far side is another huge sculptural figure, also horrifically distorted; a human form apparently disintegrating into black ribbons that stream out behind as it desperately rushes forward.

Figure 6.4 *Casting a Dark Democracy*, Tim Shaw, Aberystwyth Arts Centre (2014), Image courtesy of the artist; anonymised by author.

These figures loom oppressively as they are approached. On closer examination their skeletal form becomes clear, being made up of metal framework and barbed wire draped in strips of black plastic sheeting. Cables coil from the hooded figure down and across the sand. After a time, it becomes hard to stay in the space any longer. Breathing becomes unpleasant and the noise is too much. Upon exiting the space there is a sense of relief, but the repetitive noise continues to be audible out into the gallery and the taste and smell of oil linger.

Black Smoke Rising staged an event with multiple relationships to its own elements, the Iraq war and other events. Each of the main elements of *Black Smoke Rising* has been realised across several group and individual exhibitions, but not only have these works been realised slightly differently in each case, their relationships to events have also altered over time as the artist has amended them. Tracing this process draws out the ways in which artworks can be interpreted as evental assemblages and events as multiple.

In the first iteration of the hooded figure, presented at the Kenneth Armitage Foundation in Kensington in London in 2007 and 2008, the edges of the pool of oil reproduced the outline of the figure, so that it resembled a shadow, while also reflecting it. But as Shaw has related,

> when I originally worked on *Casting a Dark Democracy*, it was of vital importance to have the shadow of the figure cast in oil across the floor, because I was making a political statement about the UK's decision to go to Iraq. Whereas now [2014], this figure transcends that war. And so creating this rectangular pool instead transforms the work into a memorial. (Khanna and Shaw 2015, p.21)

While the form of the work might therefore be said to have shifted from 'intervention' to 'memorial', it is worth noting that during *Black Smoke Rising* the Chilcot-led Iraq Inquiry was still taking place in its effort to construct an authoritative account of the war from the perspective of British governmental decision making. By enacting the event in a different way, in parallel with the inquiry, *Black Smoke Rising* served as an ongoing counter-actualisation of the event that Chilcot was also investigating.

In the artwork, the hooded figure is 17 feet tall, far larger than a human body, and, standing on a box, it looms over the space in intimidating fashion. Shaw recalls his first encounter with the photograph of the figure at Abu Ghraib, which he describes as a shock:

> That image was overwhelming for me, it was visceral, in that it made my very insides turn for a long time… I looked at it, I tore it off the rest of the paper and I put it on the wall that morning in the studio and it didn't leave my studio until I took it London where I worked as well, and always had it. That morning when I looked at it, I said I would make a piece of work about it. (interview with author)

As Shaw describes, then, the image of the man on a box was experienced as an affecting event, but one in which he distinguished the image from the person in the picture (in Khanna and Shaw 2015, p.21).[14] Shaw further recalled the sense-making processes elicited by the image, in the following terms:

> it looked like something that was dug out of the ground, out of the sands of the ancient past; dug out of the human collective consciousness. As I examined it and worked about it, what I felt about that image is that it was the shape and the form of man's darkest, deepest, innermost collective fear. It also has all these different resonances and references, to the Ku Klux Klan, the Santa Semana, Holy Processions and other things… Strangely, not long before I was a resident artist and I'd travelled across many of the ancient sites, and in the Athens archaeological museum there was a small bronze figurine of a cloaked, hooded person similar in appearance to the Abu Ghraib torture prisoner… I remember looking at that, it looked curious, quite dark. I did a small drawing of it in my book, just a very quick drawing… So you can imagine, months later, seeing that image and thinking about the past, and also having been brought up in the early seventies in Belfast, it was a collision of feelings that came to me. (interview with author)

For Shaw, then, religious iconography forms a reference point for understanding and responding to this image, and this further serves to disengage it from the actual person shown in the photo.[15] The image is understood on an affective and virtual rather than personal level, subsisting 'as something that trawls just beneath the surface of the collective consciousness, revealing to us inherent primitive instincts, and belongs to no particular age or place' (Khanna and Shaw 2015, p.21). Shaw also recalled that priests entering the presence of a large sculpture of Athena in the Parthenon would 'quake in their boots... because they believed that the spirit of Athena was within the form' (interview with author). As he relates, 'with *Casting...*, I think there was an element that I wanted to get within that of this inner sanctum, that when you walked in there, there was a religious aspect to it' (interview with author).

Another formative influence on the works presented together in *Black Smoke Rising* are Shaw's experiences growing up as a Protestant in Belfast during the Troubles. As he has described, 'We all grew up with barbed wire ... there were always a lot of barriers and roadblocks everywhere in the 70s' (Khanna and Shaw 2015, p.21). As he relates, the Abu Ghraib image in particular 'reminded me a lot of Northern Ireland and what went on. You become very aware of what people can do' (interview with author). At the same time, it is not a question of channelling these events directly into the work; there is a question of how temporality as well as spatiality themselves are encountered from a particular subject position and within a particular community of sense. It is only with the passage of time and the changing political situation – that is, with the coalescence of a particular structure of feeling – that it has become possible to open up to some extent about past events:

> people don't like talking about the Troubles, still... It's an odd thing, because I as an artist, it's probably shaped a lot of what I've done, whether it's an embarrassment or something that we just don't want to talk about... Definitely it was a no-no in Northern Ireland and certainly you wouldn't want to be talking about it in a bar. It was just a habit that you acquired, that you just didn't... I can talk about it now, but as I say it's taken years to process it. Why it happened, what happened, I didn't live in the Falls Road or the Shankill Road where it would've been really bad. But still everybody was affected. (interview with author)

The relationship, but also distance, between events in Iraq and Northern Ireland was brought home to Shaw in a further event, an encounter with two Iraqi people who brought friends to see *Casting a Dark Democracy* at the Kenneth Armitage Foundation. As he relates:

> They gave me a real insight into the situation, because I've never been to Baghdad nor Iraq... One of them said, 'You just do not know what it is like, when an army come in from another part of the world.' And I don't know what it would have been

like. I suppose if I had been a Catholic living on the Falls Road I might have known that feeling, of somebody coming in from another place and taking control. (interview with author)

Just as the Iraq war is widely understood as a war for oil, the figures in *Black Smoke Rising* are deeply implicated with oil, coloniality, modernity and the anthropocene. The metal cables that snake out from the hooded figure flow into the sandy ground, suggesting some subterranean relationship with the pool of oil in front; curator Indra Khanna (in Khanna and Shaw 2015, p.21) has suggested that 'the whole figure seems constructed of the materials of its own torture'.

It is further possible to broaden this reading to consider how the installation employs, and alludes to, processes such as combustion that release energy, transform bodies and emit new toxic substances. These processes, affects and effects are enacted in the sculptural figure that appears to be enveloped in flames, *Man on Fire* (Figure 6.5). The immediate prompt for this work was a widely published photo of a British soldier escaping from a Warrior vehicle that had been engulfed in fire from a petrol bomb in Basra in September 2005. The petrol bomb had been thrown from a crowd that had gathered to protest an operation by British forces to extract two SAS soldiers from an Iraqi jail, after they had been detained when failing to stop at a checkpoint in the city. The soldier who is pictured

Figure 6.5 *Man on Fire*, Tim Shaw, Aberystwyth Arts Centre (2014). Image courtesy of the artist.

escaping from the vehicle in flames was one of three hurt in the incident; two civilians were reportedly killed and 15 injured (BBC News 2005a). The combustive nature of this incident resonated with an event that Shaw describes having encountered in Northern Ireland around the same time:

> I drove my van into a riot scene, about a hundred, two hundred yards away, stupidly. It was around 2005 and there had been riots taking place, I think it was something to do with too many concessions being given to the Republican side. I was returning from Enniskillen, from the western side. I could see the black smoke and I knew the nature of that smoke, heavy black smoke from burning tyres. I thought, 'I must go there' ... and I arrived along this road with cars burnt out and the tarmac boiling with anger, and this came into the work as well. And then it's about a hundred people in front, about a hundred yards away and I realised, 'What the fuck am I doing in this place?' I just turned and drove away rapidly. (interview with author)

The work is further connected with still other violent events, notably the bombing of a department store in Belfast in 1972, which Shaw experienced as a seven year old shopping with his mother (interview with author; Jordan 2015). As well as working with specific materials, then, *Black Smoke Rising* appropriates and seeks, in a process of counter-actualisation, to transmute the violent, energetic transformation of bodies and things in the Iraq war and well beyond.

Conclusion

Souvenir from the Ministry of Justice, 20:50 and *Black Smoke Rising* offer distinct enactments of oil in relation to the war that are not just affecting, but which foster, I argue, a genealogical and virtual sense of an event taking place across multiple spaces and times, and intersecting with other events in multiple ways. They are able to operate in this way, I suggest, because they are evental assemblages, arrangements of material and symbolic elements that can be understood as having been brought together out of events, as themselves constituting events, and as engendering further events when activated by viewer/participants.

Souvenir from the Ministry of Justice might be seen as a polemical retort to a war that was transparently 'for' oil, but can also, I have suggested, be interpreted as modelling the war genealogically in relation to the formation of the Iraqi state as well as the sites and institutions of international law. In drawing attention to the enforcement of the rule of law in The Hague, it further highlights the impunity with which the Iraq war was conducted. If one of the critiques of artworks responding to geopolitical problems (e.g. Graham 2010) has been that their ephemerality and isolation from social movements limits any political effect they might have, it is also possible to turn this around to argue that it is precisely in virtue of their limited spatio-temporal duration and quasi-autonomous character that they are able to engage events in an affecting manner, by constituting a temporary present or 'specific sensorium' (Rancière 2002, p.134, fn.1) in which the world is configured in a different way. Their apparent limits are a condition for them being the kind of events that they may be encountered as being.

This point is illustrated differently by *20:50*, which uses oil in a way that is isolated from any direct reference to political conflict or environmental harm. The reflective nature of the surface of the oil means that it is even possible to encounter the work, if only briefly, as not involving oil at all. To the extent that it might exert a critical effect in London, *20:50* might be said to have operated indirectly, via the critical knowledge and imagination the viewer might bring to the work, in that they might, for example, have connected its presence at the Saatchi Galleries with London's role in the geopolitics of oil. The conjunction of the sensory and the geopolitical is much harder to overlook in case of the installation of the work in the Red Jail, while the simultaneous existence of the work there and in London establishes a temporary link between them. *20:50* can therefore be thought of as existing via least three distinct events: the version in London (in fact four different versions), the version in Sulaymaniyah, and the dual existence of the work in both places at the same time. These instances are part of the events of contemporary art and of Wilson's own practice, but also intersect with multiple geopolitical events including Iraq's formation as a state, Kurdish struggle and the anthropocene.

Black Smoke Rising connects sensation and sense making in different ways. Rather than fostering a moment of contemplation in the manner of *20:50*, the viewer enters into an environment saturated by oil and by the violent material energetic transformations that it enacts and drives. The experience of the work is more directly of violent events: of the violence of the Iraq war and of specific incidents taking place within it, but it is also in touch with the violent geographies of the British state and its counter-nationalisms, as well, again, as the events of petroculture and of the anthropocene. *Black Smoke Rising* is also suggestive here in relation to work that approaches the anthropocene through fire and combustion. As Steve Pyne (2009, p.443) has written, the anthropocene 'is essentially an era of anthropogenic fire... even global warming is largely an outcome of humanity's combustion habits'. As Pyne (2012, p.120) has also observed, while artistic interest in fire has come and gone in Western art history, the burning city is an 'enduring theme'. As the preceding, current and subsequent chapters explore, fire, like oil, has surfaced repeatedly as both theme and co-producer of artworks emerging from the Iraq war. While we might make sense of the installation through the well-established concepts of geopower and biopower, it also calls forth an engagement with the event of what Simon Dalby (2017) has called firepower.

Notes

1   I discuss other works by Rashad Selim in Chapter Eight.

2   The first recorded Western traveller in Mesopotamia was Benjamin of Tudela, who voyaged from Spain through the region to China between 1162 and 1173, visiting Babylon (Bernhardsson 2005).

3   Such speculative thinking was further in evidence in a discussion at the Royal Geographical Society in January 1910, in which Maunsell (now a Colonel) and Gertrude Bell also participated and which was led by a paper by Sir William Willcocks, who had recently been appointed as an advisor to the Ottoman government on the irrigation of Mesopotamia (Brown *et al.* 1910).

4   As Sluglett writes, 'the Empire air route, the oilfields, the RAF training ground, British prestige and investments ... could not be given up simply because the Iraqis did not want them' (2007, p.63).

5   I am grateful to Rashad Selim for alerting me to the existence of these works in the IWM collection.

6   Four British steamships of the India navy, named *Assyria, Nimrud, Nitocris* and *Euphrates*, sailed into Mesopotamia in the 1840s (CR Low 1887, cited in Saleh 1966: 203).

7   The work has also been installed at the Royal Scottish Academy in Edinburgh in 1987, at the Mito Art Tower, Japan, and at the Australian National Gallery in Canberra in 1996 (Morrissey 2005). In late 2018 it reappeared as part of the *Space Shifters* exhibition at the Hayward Gallery on the South Bank in London.

8   See http://artrole.org/projects/post-war-festival/ (accessed 4 July 2017).

9   Here I rely on photographs and descriptions of the work and festival available online and an interview with Adalet Garmiany, artist and curator as well as founder and director of ArtRole. Garmiany studied at the Institute of Fine Art in Mosul and later the University of Lincoln and Hull School of Art and Design in the UK. He lost many family members in Saddam's 1988 campaign against the Kurds and left the country in 2000, returning after the 2003 invasion. Interview with Adalet Garmiany, 24 February 2012; also https://artsfacultyresearch.wordpress.com/2013/04/22/mark-terry/ (accessed 4 July 2017). Wilson has also stated that he felt the work had 'gone to its spiritual home, given the amount of oil Iraq has/had' (email correspondence with author).

10   See http://artrole.org/projects/post-war-festival/ (accessed 6 December 2017). 4 July 2017. Garmiany stated that 'our main mission is to build a bridge between UK, Iraq, the Middle East and the United States as well' (Perks *et al.*, 2012).

11   The work was curated by Indra Khanna and produced in collaboration with MAC Birmingham and Aberystwyth Arts Centre with support from Arts Council England in summer and winter 2014.

12   Photographed by the author. Warnings to visitors concerning the hazards posed by oil were also posted at the installations of *20:50* at City Hall and the Saatchi Gallery in London.

13   This element of the exhibition, *Soul Snatcher Possession*, presented in a claustrophobic room constructed from plywood in one corner of the space, staged and explored issues of embodied violence. Here several sculpted figures, masculine in form, seem

to be caught in the midst of an act of violence, perhaps a punishment beating, on the edges of which are positioned an apparently male figure suggestive of authority and an apparently female figure, which may also have experienced an act of violence and which may or may not be related to the central event apparently still taking place (Khanna and Shaw 2015, p.26).

14   The person pictured in the original photograph has never been publicly identified in a conclusive way (see summary in Hudson 2015). Hudson thus frames the work as 'a monument to unknown prisoners *everywhere*' (2015, p.18). As he writes, in 'breaking away from its documentary and experiential beginnings, Shaw's work not only gains strength, but also achieves universality' (Hudson 2015, p.19).

15   In a critical essay on the work, Hudson writes, 'in the curious, beckoning image of the arms, seen mirrored in the pool of black crude oil at the figure's feet, is a kind of ghostly after-image of the Saviour' (Hudson 2015, p.9). As the critic Alison Kinney (2006) has argued, the hood is a garment that is often bound up with the exercise of power, in a wide variety of historical, geographical, cultural and political circumstances and events. In the terms of this book, 'the hood' is virtual while particular hoods are actual.

Chapter Seven
Photomontage as Geopolitical Form

Photo Op, a 2005 photomontage artwork by kennardphillipps (Figure 7.1) that appears to show a grinning Tony Blair taking a selfie against the backdrop of a burning oilfield in Iraq, has come to function as a definitive image in relation to Britain's 2003 Iraq war and to contemporary geopolitics more broadly. This work has not only been accorded this central status by commentators (e.g. Jonathan Jones 2013), but, again and again, when writers, editors and curators have selected an artwork to illustrate an article, decorate a book cover or promote an exhibition having to do with the war or related issues, it is *Photo Op* that they have selected. As well as being included in numerous exhibitions, *Photo Op* has also been adopted by protestors, and several versions of the work have been acquired by major public institutions. The work has become so recognisable that it has also come to serve as a template for other artists and designers to borrow and modify for their own purposes. But how could an imaginary image, which shows an event that did not happen, come to play such a role in relation to an event that did, and what does this tell us about how art – and more specifically, photomontage – might work in relation to geopolitical events? As it happens, *Photo Op* is also the title of a 2004 photomontage created by the American artist Martha Rosler in relation to the interventions in Afghanistan and Iraq. But while Rosler's work has been recognised by writers on geopolitics, discussion has remained circumscribed, overlooking the genealogy

Geopolitics and the Event: Rethinking Britain's Iraq War Through Art, First Edition. Alan Ingram.
© 2019 Royal Geographical Society (with the Institute of British Geographers).
Published 2019 by John Wiley & Sons Ltd.

Figure 7.1   *Photo Op*, kennardphillipps (2005). Image courtesy of the artists.

of photomontage as form and the complex ways in which photomontages emerge from, enact and relate to events.

In this chapter, I rethink photomontage as geopolitical form through a more detailed consideration of the work of Rosler, and build on this to discuss the work of kennardphillipps and to explore the relation of kennardphillipps' *Photo Op* to Britain's Iraq war. In the first part of the chapter, I critique the reception of Rosler's work by writers on geopolitics, placing it more fully in art-historical context and drawing more fully on the artist's own account of her practice. This discussion allows a fuller sense of the multiplicity of photomontage as a form of evental assemblage to emerge. I then bring this approach to bear on the pre-Iraq war photomontage work of Peter Kennard. The chapter then turns to the work of kennardphillipps, the collaboration between Kennard and Cat Phillipps since 2002, and presents an account of the creation, form, circulation, reception and appropriation of their *Photo Op* (2005), exploring its emergence as a definitive work in partial contrast to another work made around the same time, *Know Your Enemy* (2005).

Towards a Genealogy of Photomontage

Recent discussions of photomontage in political geography and international relations have tended to focus on the work of Martha Rosler, and in particular on two series of works, *House Beautiful: Bringing the War Home* (c. 1967–1972), which dealt with the Vietnam War, and *House Beautiful: Bringing the War Home, New Series* (2004–2008), which addressed the wars in Afghanistan and Iraq.[1] In both series, elements of photographs taken in Vietnam and in Afghanistan and Iraq are juxtaposed with images from American lifestyle magazines, in ways that highlighted the contrasts between them, while suggesting that they might in some way be related.

Commentators in political geography and international relations have readily detected some of the geopolitical dimensions of these works. As Derek Gregory (2010a, p.177) has suggested, the *New Series* compositions 'show the Iraq war erupting into the American home...'. As he also argues,

> Rosler's sharper point is to goad her audience... to recognise how often 'our' wars violate 'their' space: her work compels us to see that what she makes seem so shocking in 'our' space is all too terrifyingly normal in 'theirs'. (2010a, p.177)

Katharine Brickell (2012, p.575), recognising their specifically feminist orientation, notes that Rosler's works 'shatter any illusions we may hold of geopolitics and home being somehow separable'. As Michael Shapiro (2012, p.146) remarks, 'the juxtapositions of the commodity-saturated interiors with war scenes... have an unsettling effect that must engender reflection on the ways in which everyday life is politically insulating'. Each writer, then, finds Rosler's work affective – compelling, shattering and unsettling – and *effective*, in that it elicits enhanced critical awareness of ongoing events. These works are thus understood as offering counter-events that disrupt and reassemble the world, sensation and sense-making.

Although Rosler's work has been recognised by writers interested in geopolitics, discussion remains somewhat abbreviated and has taken little account of Rosler's own accounts of her practice.[2] As she has explained, both series were indeed made with explicitly critical geopolitical intent, to break with a sense of the world as it was being given:

> I was very much interested in the picture of the world that our culture propagated, one that suggests that there were numerous 'worlds', none of which quite intersected with one another or, if so, in some arcane way. I wanted to suggest the unity of the world and therefore our – at least putative – responsibility for what went on within it, and that in particular related to the space of representation in which women were inserted, and which we disclaimed as being actually about us. Also of course it was to do with the notion of a war elsewhere, on others' territory, which we could say was happening 'over there', 'outside somewhere' although we could hardly disclaim responsibility. So that was my aim. (Martha Rosler interviewed by Iwona Blazwick 2008, p.3)[3]

Rosler's Vietnam era and post-2001 works employ photomontage techniques to question the compartmentalisation of time, space and events that she saw as prevailing in US culture and to intensify an alternative structure of feeling. In focusing on the quality of shock or the jarring juxtaposition between elements that seems to be evident in Rosler's work, however, commentators have overlooked the extent to which they also rely on recognition and other sense-making capacities on the part of their viewers. Again, as Rosler (2004, p.355) has explained, with explicit reference to spatiality: '[i]n all these works, it was important that the space itself appear rational and possible; this was my version of this world picture as a coherent space—"a place"'. Also overlooked has been the role of narrative in sense-making practices. Rosler describes her turn away from sculpture and abstract expressionist painting towards photomontage in the late 1960s as being motivated by a desire to work with 'an imaginary space *in which different tales collided*' (2004, p.353, emphasis added).[4] Something else that has generally been overlooked is the symbolic violence that photomontage must effect in pursuit of its critical goals. Both series of *Bringing the War Home* include works containing images of violent acts, distressed people and wounded children that have been removed from their original photographic contexts, and combine them with normalised or idealised representations of American women and domestic spaces, but little attention has been given to the stakes or implications of this tactic.[5] Rather, its critical effect tends to be assumed or asserted, rather than explored or explicated (see also Ingram 2016).

Consideration of the specifically evental qualities of photomontage also remains circumscribed. Shapiro argues that Rosler's photomontages involve the juxtaposition of two events: American intervention and the commoditisation of domestic life, eliciting a third event of shock and critical apprehension within the viewer. However, closer attention to Rosler's *Photo Op* (2004), (Figure 7.2), which provides the cover image for Shapiro's *Studies in Transdisciplinary Method* (2012), reveals that at least six distinct events have been brought together by the artist in a single frame. The work therefore bears more detailed consideration than it receives; each of these events could be further unpacked and explored, as could their careful, artful composition.

We can further expand our understanding of photomontage in relation to the event not just by considering production and form, but also via a more extensive consideration of circulation, reception and appropriation. Rosler originally sought to avoid her Vietnam series becoming artworks. As she has said, 'I thought of these works as "not art" and refused invitations to show them in art institutions' (Paco Barragán interview with Rosler, 2012), and she accordingly aimed to disseminate them outside of the gallery and the auction room, but rather through activist networks: their 'site seemed more properly "the street" or the underground press, where such material could help marshal the troops' (Rosler 2004, p.355).[6] Over time, however, they came to be recognised as artworks both in relation to Rosler's own identity as an artist and via their reception in artworld contexts.

Figure 7.2   *Photo Op*, Martha Rosler, from the series *House Beautiful: Bringing the War Home, New Series* (2004). Photomontage © Martha Rosler. Courtesy of the artist and Mitchell-Innes & Nash, NY.

The geopolitical import of the 2004–2008 series, meanwhile, derives in part from its art-historical relation to Rosler's previous work:

> I have consciously chosen, in effect, to quote my own method of work from forty years previously in order to create a meta-level commentary about the failure of our political class to learn anything from history. Today, we have new wars of choice that are being waged with the old mindset, so I chose to use the same mode of address: the photomontage. (Bree Hughes interview with Martha Rosler 2013)

In Rosler's work, the event of the Iraq war is thus tied to the Vietnam War, but the digital reproduction and dissemination of the latter work, as well as its more direct address to the artworld, over and above its content, registers differences between the two events, as well as repetitions and rhymes.[7]

There is also the question of viewership and positionality. Rosler has stated in relation to her adoption of photomontage that

> I was very taken by the idea of giving a viewer a place to stand, and therefore the photographic became the obvious choice because photography tends to suggest the possibility of a real space, if you don't just cut it up and ignore the idea of perspectival relationships. (Martha Rosler interviewed by Iwona Blazwick 2008, p.3)

Here the technical details of composition are crucial to what Rosler's photomontages are understood to be and how they might be said to work. While the early photomontages created by Dada artists after the First World War exploded space and time, the famous anti-Nazi works later created by John Heartfield aim for much more direct and legible political communication. In Rosler's works, the unreal and real are aligned to produce spaces that are at least plausible, such that the kinds of narratives one might intuit are more or less within reach.[8] It can further be inferred that their viewers are imagined to be at-least-potentially concerned American citizens, in America: these works are composed not just in, but for, a particular time, place, audience and purpose. Vietnamese, Afghan or Iraqi people would probably have little need of being enlightened or compelled by such a work. Photomontages, then, are also situated with regard to their imagined viewership and reception.

As I have argued more generally, it is useful to consider how artworks emerge out of events, how they work as events, and how they may become entrained in further events. A consideration of Rosler's photomontages pushes us to open out this heuristic, however, first, because its initial relation to events is already complex: photomontages work with things that are themselves both events and representations of events. These photo-events are then manipulated in the further event of assembling the montage, something that takes place amidst the artist or artists' encounter with the geopolitical event more generally. The event of the work may also itself be more complex than the collision of two elements, and the ways in which it works as an assemblage may itself vary depending on how it is exhibited, used or encountered (Dittmer 2013). A photomontage may then be further circulated, used, quoted or modified, and each of these events may somehow flow back into the larger geopolitical event, raising new questions of ethics and positionality. Our understanding of photomontage as an evental geopolitical form can be extended by conceptualising and seeking to investigate each of these moments.

Photomontage and Geopolitical Events

The question of the event forms a recurring concern in the photomontage practice of Peter Kennard and more latterly kennardphillipps.[9] Kennard trained at Byam Shaw, the Slade and the Royal College of Art but reacted against the way in which art was being taught in the late 1960s, which seemed to bear little relation to ongoing political and geopolitical events. In this context, Kennard developed an interest in techniques that could engage images of events, while in some way also evoking their violence. One of the techniques with which Kennard experimented in early work was drypoint, in which lines are cut into a metal plate, leaving a burr along the edge of the incision, lending a characteristic fuzziness to the image that is then printed. As he has written,

[t]he burr of the drypoint line, in registering the violence of its own making, was a symbol of my desire to express through the human subject the brutal physical attacks that I knew were happening, especially on Vietnam. That sense of ripping into an image, unveiling a surface, going through that surface into a revealed truth, is [also] at the core of photomontage. (2000, p.35)

While expressed in terms of revelation, here Kennard also comes close to Deleuze's (2015) account of how, via a process that doubles its violence, an embodied artistic performance may detach an event from its self-evidence. In Deleuze's account, what happens is less revelation than a movement towards the virtual, or a counter-actualisation, which might then be actualised differently.

Kennard's turn towards more forceful techniques and to photomontage in particular was informed by his discovery of the work of Dada artists John Heartfield and Hannah Höch and the Russian constructivist Alexander Rodchenko. Critical art practices forged in Weimar Germany and early Soviet Russia (Willett 1978) were thus appropriated and redeployed in the midst of Cold War conflict. Kennard's account of his practice around this time emphasises its relations to events and to the world, in ways that further chime with immanentist and genealogical thinking. As he has written: '[a]lthough compared to a photojournalist, I'm grounded and stuck at my safe desk or in my darkroom under safelight, I still feel that my subject has to be made up from events around the world' (Kennard 2000, p.44). This involves grappling with the question of one's connections to events that are not encountered directly: 'I have not experienced the horror of Kosovo or East Timor, but I am a citizen whose actions are globally tied up with what's happening in the world. ... To me it's about making work as a citizen' (Kennard 2000, p.42). This leads to a rejection of the idea of art work as intervention, as this 'suggests being outside of the experience that you depict' (Kennard 2000, p.42). As he has also related,

[m]y photomontages are constructed to show that events are not as natural or inexorable as they often appear to be in photographs, but the result of choices and decisions. They aim to subject photographic images to human criticism. (Kennard 1990, unpaginated)

Kennard (2000, p.46) has further described the aim of his work in evental terms, as using iconic images in order 'to render them unacceptable', and '[a]fter breaking them, to show new possibilities emerging in the cracks and splintered fragments of the old reality'. He has also rejected the idea that the flow of images is necessarily overwhelming and has asserted that the artist must be 'an active critic' (Kennard 1990, unpaginated). His practice can thus in many respects be recognised as evental, and as offering not just a critique, but a counter-actualisation, in that his works seek to break the self-evidence of events and to make something else of them.

Kennard and kennardphillipps' work can also be contextualised with regard to the institutions, discourses and practices of the art world, or as I have conceptualised it, the *dispositif* of art. Kennard has been attracted to the idea that photomontages 'exist through their reproduction rather than as valuable commodities' (2000, p.42) and he has sought to embed his practice and make it work more in the context of political struggles than the art world, with gallery and museum exhibitions often gathering works that have already been used in a variety of campaigns and newspapers rather than, or at least alongside, works made specifically for an institution. By 2000, Kennard had made, by his own estimation, some 600 photomontages, many of which were distributed by left-wing publications, including *Workers Press, Socialist Worker, New Statesman* and *The Guardian* as well as for social and political campaigns, including the Campaign for Nuclear Disarmament (CND). While many of kennardphillipps' photomontages can now be bought in fine art form, they have refused payment for work they see as being made as part of campaigns and struggles. Their works are made in order to be readily reproducible and they have encouraged people to adopt and appropriate their images, offering works as free downloads and running workshops teaching montage techniques to people who do not have formal art training.

Photomontage was Kennard's favoured technique for some two decades, but in the late 1980s, he initiated a second strand of work focused more around objects than images, citing '[a] mixture of personal experience, disillusion with organised politics and the use by media of innumerable digital photomontages', which caused him 'to question the effectiveness of photomontage as a critical, social probe' (Kennard 2000, p.35). While he continued to make photomontage works responding to specific campaigns and events, including the 1991 Gulf War, this period is marked by a questioning of the efficacy of the form in light of digital reproduction.[10]

Kennard's pre-2002 photomontage work appropriated the radical political aesthetics developed in inter-war Europe, but these works tended to avoid the kinds of complexity evident in some Dada and post-Dada photomontages (Kriebel 2014) or in some of Rosler's Vietnam and war-on-terror works, evoking more closely John Heartfield's later anti-Nazi montages, which aimed at popular political communication. His photomontage practice has been grounded furthermore not just in the wish to break the power of ideologically-laden images, but to make the resultant works impossible to reappropriate, and to foster a critical attitude towards images among a broad spectrum of people. Indeed, Kennard (2000), and since 2002, kennardphillipps (interview with author), strongly question the value attached to art works that are ambiguous, obscure or which otherwise strive for autonomy. At the same time, there is a recognition that art politics are distinct from formal politics; Kennard has accordingly avoided membership of political organisations. Furthermore, while some radical artists have opted out of the gallery and museum world altogether, seeing it as beyond redemption, Kennard and kennardphillipps have allowed their work to be acquired, curated and exhibited, holding it to be important for critical imagery to be present in official institutions in order to become accessible to the public and to oppose the work that appears there (interview with author). kennardphillipps also attach importance to the opportunities afforded by museum and gallery education programmes to train people in critical art techniques.

This brief overview further reinforces the necessity of interpreting photomontage in terms of a cluster of practices and processes – production, composition, circulation, reception, and appropriation – which can in turn be conceptualised as events taking place amidst and in relation to events, of which they form parts, but which they can also be understood as counter-actualising.

Kennardphillipps and Britain's Iraq War

Kennardphillipps formed in 2002 in anticipation of the invasion of Iraq, but did not create new work in relation to the event until 2004. In the meantime, Kennard's montage *Defended to Death* (Kennard 2015) was reworked as *Union Mask* and adopted for use in protests by the Stop the War Coalition, enacting a link between Cold War and war-on-terror era political aesthetics. A banner version of *Union Mask* can be seen draped over the head of one of the lions at the foot of Nelson's Column in Trafalgar Square during a protest against George W. Bush's visit to London in November 2003.[11] Somewhat untypically, the first work created jointly by Kennard and Phillipps, a printer with particular expertise in digital techniques, was oriented towards the fine art world rather than activism and protest. *Award* (2004) was a series of fine art prints that also referred back, to montages created by Kennard in opposition to the Falklands war in the 1980s and to other critical artistic appropriations of the medal form (Attwood and Powell 2009) (Figure 7.3).

This series reworked the motif of the war medal using distressed ribbons, with the medals degraded or replaced by images including a military helicopter, a targeting reticule and a just-discernible burnt body. These works were made by assembling objects and materials (including medals, bandages, dust, oil, grit, photographs and blood) on a digital scanner, which was used to create images that were then combined to produce the final compositions. The process of making these works again illustrates how art history, materials and the event intersect in the event of practice:

> As the Occupation became filthier our prints became denser and denser. The scanner became like an etching plate and through using the newest technology our practice became strangely more and more connected to printmakers of the past. We looked at Rembrandt and Goya etchings, at how the etching, aquatint and drypoint layers on their metal plates became our dust and fabric, set layer upon layer on the computer.[12]

Figure 7.3   *Award*, kennardphillipps (2004). Image courtesy of the artists.

The artists sought to translate events into images by means of objects, materials and practices, in order to engender new events.

The *Award* works were first exhibited in 2004 at the Henry Peacock Gallery in Bloomsbury, and were shown again as part of *Print Now: Direction and Definitions* at the Victoria and Albert Museum in 2006.[13] However, the experience of making and exhibiting these works in relation to ongoing events in Iraq led to a shift in the artists' practice:

> We thought they were too easy on the eye, which is why we changed tack after that. They weren't heavy enough, they were too decorative in a sense, as images, which is why we started doing the more direct montages, the posters. Although it's never as simple as either-or, but especially when you're working on something about the horrors going on 'over there', you always feel that what one does doesn't totally match up, so one tries something different. (Peter Kennard, interview with author)

The turn to photomontage thus emerged out of experimentation in relation to the event. While kennardphillipps continued to create works and installations using a variety of media and techniques, photomontage acquired a renewed centrality in their practice. Here I focus on two photomontages that they created in 2005: *Know Your Enemy* and then *Photo Op*. While both illustrate the multiple relations between photomontage and the event, it is the latter work that has gone on to become near-ubiquitous.

Know Your Enemy

The evental processes and practices involved in kennardphillipps' photomontage work are well illustrated by the work *Know Your Enemy*, created as part of the *STOP* series of posters that was produced *in situ* at *War on War Room* at EastInternational 05, curated by Gustav Metzger. In this project, photomontage posters were reproduced on newsprint and distributed for free, while the artists' equipment was made available for visitors to create their own works. This project thereby embodied the artists' tactic of making work within, and for, the public domain; during their residency, kennardphillipps also carried out an action where they sought to attach the posters to the fence surrounding the US Air Force Base at Lakenheath in East Anglia, a move that was stopped by the police. This incident was captured on video, which was in turn was exhibited in an art context.[14]

Know Your Enemy is an appropriation and reassembly of two images that had appeared widely in the British press (Figure 7.4). The background to the work is provided by a press agency shot of Tony Blair and George W. Bush entering 10 Downing Street. In the photo, Blair and Bush are in the process of stepping over the threshold and Bush is reaching across, with his left hand resting on Blair's left shoulder, guiding him into the doorway. The pose, from which a gesture is readily inferred, suggests a desire to assert superiority. What also captures attention, though, is a scene in the foreground, in which a soldier in a dark British army uniform is apparently caught in the middle of a punch to the head of a prone male figure wearing only a vest and shorts, whose upper body is tied up in blue webbing. The figure lies in a posture that is immediately recognisable as passive and protective, perhaps having already been hit or anticipating the blow that appears to be on its way. Again, the eye readily infers animate motion, with intent apparent in the posture and facial expression of the soldier. One complex event – the abuse of detainees – is captured and set against another – the junior-senior relationship between Blair and Bush – amidst the larger event of which they are each part. The montage of the two appropriates each of these events, constituting a further counter-event. The association and juxtaposition between the two images is achieved, however, via the meticulous, skilful selection, composition

Figure 7.4   *Know Your Enemy*, kennardphillipps (2005). Image courtesy of the artists.

and assembly of each of the elements, and the resulting work both invites and repays closer consideration.

While diplomacy in London and the violence of occupation in Iraq are topographically separate, the montage places them in the same spatio-temporal framework, with the alignment of perspective, scale, lighting and resolution between the two images making them initially appear to be a single 'real' event. It seems obvious from the montage that Blair is not just the junior partner, but is blithely ignoring blatantly illegal violence committed by the armed forces that he commands. The montage does not so much reveal abuse or a link between political leadership and embodied violence, as dramatise it in ways that can elicit narratives via the play between visual media, a viewer's embodied perceptual capacities (which enable an apprehension of movement, violence and response) and their knowledge about ongoing events (which may or may not enable inferences and storylines to emerge). Affect or feeling does not necessarily (or at least only) precede cognition or thinking; the recognition of what is 'happening' in the 'event' can intensify one's affective response to the image;

the encounter with the work is itself a complex thinking/feeling event that is intended to intensify opposition to the war.

Our evental understanding of the work is expanded further by considering in a little more detail the provenance of the images and the events of which they are a trace. The image fragment in the foreground is taken from a photo that entered the public domain in January 2005, when, along with 21 others, it became part of the evidence in court martial proceedings against three British soldiers over what was described as the 'abuse' (rather than torture) of Iraqi men who had been detained at a location called Camp Breadbasket, in Basra.[15] The photos, which had been taken in May 2003 and which only surfaced after the (pre-digital) camera film had been taken to a shop to be developed, recorded violence against men detained in what was described as an effort to deter looters from stealing from the camp, which had been tasked with distributing humanitarian aid (Oliver 2005). The soldiers claimed they had been ordered to catch looters and 'work them hard', an allegation denied by their superior officer (Oliver 2005). The pictures recorded soldiers treating the men in a humiliating manner, including being suspended by webbing from a forklift truck, in the process of apparently being kicked, and being posed in a sexualised manner (BBC News 2005c). While some of the charges were dropped, the men were dismissed with disgrace from the armed forces and sentenced to jail terms, with the Judge Advocate particularly criticising the posing of the men for the purpose of taking trophy photos.

The background and middle ground are provided by a photo taken during George W. Bush's visit to London in November 2003, an event that took place after the prisoner abuse had happened, but before it became public, and which was marked by major protests in the city. The shot in the montage corresponds closely to a press photo accessible online, which appears to have been taken a moment beforehand (Getty Images 2003). Though her presence does not suggest itself on initial viewing of the image, Laura Bush can also partially be seen, having entered just a moment before. This photo corresponds with others taken shortly beforehand, of Blair, Bush and their wives outside number 10. Though the door (and thus the number 10) is not visible, the location the arrangement of decorative window, door frame, cut stone and doorbell call up recognition of the address, a metonym for the Prime Minister and his government. In placing the two elements together, the montage creates an event that is unreal, in that it does not depict things that happened within the spatio-temporal framework within which they are pictured, but which elicits narratives concerning a larger event in which both elements are held to be linked both causally and morally. The work reassembles events, pointing towards virtual arrangements of which they can be understood to be an actualisation.[16]

Again we may note that the manipulation of images of violence and suffering raises ethical questions, particularly in light of the work of Ariella Azoulay (2008), who has argued, following Hannah Arendt, that photos of political violence place obligations upon those who look at them to recognise the people and events within them as existing within a common world. Excising elements of such photographs is thus an act of symbolic violence, which the act of recomposition may compound. Much is therefore riding on the rationales and effects that might be said to accompany the making of photomontages about war.

These dilemmas have been recognised by the artists. As Cat Phillipps has related,

> You don't throw out the ethical element, or the pain, struggle, confrontation because if you throw it away you end up making something that is really flippant, using imagery of a moment and situation that is so grave and fatal... [but] I definitely struggled with using those pictures [in the STOP series], and it's part of the nature of how we started working, the digital technology, Photoshop ... it is physical, it is physically touching the bodies that are in the image. (Interview with author)

The process of composition is thus also not just a practice but itself an ethically loaded event. As Phillipps emphasises, the work is informed by 'a really determined and clear critique of the politics that put both the abuser and the victim in that position' (interview with author).

Here we might recall Susan Sontag's well-known critique of images of suffering:

> To set aside the sympathy we extend to others beset by war and murderous politics for a reflection on how our privileges are located on the same map as their suffering, and may – in ways we prefer not to imagine – be linked to their suffering ... is a task for which the painful, stirring images supply only an initial spark. (2003, pp.91–92)

If Sontag was commenting on news and documentary photography, then the promise of photomontage is of going further than an initial spark, decontextualising and recontextualising image fragments in an effort to make them work otherwise. *Know Your Enemy* seeks to implicate not just Blair, but an implied British viewer in the event of late colonial counterinsurgency. As Kennard has explained, 'it's actually bringing the politicians into the image of the destruction that they are causing. Obviously they're never there, normally. Photographs can't do that but montage is built for it' (interview with author). *Know Your Enemy* is not just a representation of violence or suffering, but a conscious, ethically fraught geopolitical act that is intended to elicit a response within a population, among whom, by 2005, the sense of the event as a 'wrong war' was intensifying.

Photo Op

While *Know Your Enemy* is exemplary of the tactics and rationalities of photomontage as geopolitical form, it is a different montage, one that avoids depictions of bodily harm, that has achieved near-ubiquity as a visual shorthand for the war. As I have suggested, kennardphillipps' *Photo Op* has come to be recognised, in a variety of contexts, and both implicitly and explicitly, as the pre-eminent artwork to have emerged from Britain's Iraq war. In order to draw out the evental qualities of this work, I again consider production, composition and circulation, and its appropriation in new contexts, further illustrating the close connections between photomontage as an evental form and an array of ongoing practices.[17]

The image of Blair taking a selfie with a mobile phone (a Nokia 6230, an early cameraphone rather than a smartphone) is extracted from a press photograph that was taken on 5 April 2005, in a context not immediately related to the war. Blair had just called a general election and the photo was taken on his first campaign stop. He had travelled by helicopter to the Weymouth and Portland National Sailing Academy on the Isle of Portland in the constituency of Dorset South, the Parliamentary seat with the smallest Labour majority. There, Blair gave a speech to a small gathering of invited Labour Party members, before meeting a group of sea cadets for a photo opportunity. As Oliver Burkeman reported for the *Guardian*, having given his speech,

> Mr Blair next vanished for several minutes, giving his aides time to shuffle a group of local sea cadets into position for an artfully constructed photo-opportunity beside the water. The shirt-sleeved prime minister then returned to shake their hands, and to learn, from a couple of them, how to take a picture with a cameraphone. (Burkeman 2005)

This, then, was the real 'photo op' event from which the central element in *Photo Op* was taken. Other press shots of the event (Getty Images 2005) show Blair cheerfully interacting with a dozen or so young white people, made up of six male teenagers in naval uniform and a younger group, mostly girls, in casual clothing. The cameraphone plays a central role in mediating their interactions, with Blair apparently learning the new skill of taking a selfie. The 'selfie' picture captures Blair from the waist up and centres him holding the phone, grinning broadly as he looks directly into its lens. He is holding the phone in his right hand; the angle from which the press photograph is taken ensures that the phone does not obscure the line of vision to Blair's face; instead it is one of the cadet's faces that is partially obscured. It looks like a sunny day; in the background the sea is blue and further away the cliffs and green fields of the Dorset coast are visible. There is light cloud in a blue sky. In the original, Blair is smiling along with other young people in what appears to be a shared moment of enjoyment. The cadets had not been told they were to meet the Prime Minister until the last moment; the younger children are more or less smiling towards the camera, while the teenage cadets appear to be thrilled. While the situation was staged, there is a spontaneous quality to the scene.

Photo Op removes much content and all of the context from the photo, selecting only Blair and the cameraphone; in this way a certain symbolic violence is done to the event of which it is a trace, in the service of a critical geopolitical purpose. Without the context, Blair's expression is more readily taken to be 'crazed' or maniacal (Evening Standard 2013). In *Photo Op*, the surrounding image is larger, and while in the press photo he dominates the frame, here Blair occupies less than half of its height, and is diminished somewhat as a consequence. The lighting in both elements is brighter than in *Know Your Enemy* and the two images blend less into each other; Blair stands out more sharply against a slightly fuzzier background. *Photo Op* thus suggests itself more readily as a digital photomontage, but one that might induce a certain reality effect in the viewer.[18]

Photo Op invites recognition of Blair's 'true' relation to the war as a gleeful warmonger, taking pride in the destruction unleashed by the invasion and occupation. By May 2005, the Hutton Inquiry into the circumstances surrounding the death in May 2003 of the weapons inspector David Kelly and the Butler Review of the pre-invasion use of intelligence had failed to establish Blair's culpability in a way that would lead to legal or political accountability. In *Lawless World*, moreover, Philippe Sands (2005) had not only underscored the position that the war was illegal in international law, but provided evidence to suggest that the government's legal adviser had improperly altered his view on its legality shortly before the invasion. Within a short period of time from the publication of the news photo, kennardphillipps had created *Photo Op*, and it began to circulate and to be adopted across a variety of sites, helping to stitch political practices together across translocal networks.

The work was created with a view to being used in protests against the meeting of the G8 at Gleneagles in Scotland in July 2005, where a printed version was carried and employed by the Clandestine Insurgent Rebel Clown Army group.[19] Here, versions of *Photo Op* were displayed on a placard and on a cardboard stand, which was photographed while being placed in front of police guarding the summit. One version of the work had a semi-transparent image of a smirking George W. Bush montaged into the smoke in the top left corner of the work, giving the appearance of a spirit playfully watching over the scene as Blair takes the photo. A modified version of the image, with the word STOP in capital letters in the bottom left hand corner, was then produced as a flyer for protests against the Defence and Security Equipment International (DSEI) arms fair at the Excel exhibition centre in London's Docklands in September 2005.

The work began to gain wider prominence in late 2006, when it was included in *Banksy's Christmas Ghetto*, a pop-up exhibition in a vacant shop unit at the east end of Oxford Street in central London, where for a time it was given the most prominent position, facing outwards through the store window onto the street. Photos of the exhibition taken from the street show visitors looking and pointing at the work, taking their own photos of it and glancing towards it as they walk past. These snapshots provide a small insight into the work's capacity to capture the gaze and be drawn into a variety of visual practices[20] and the November 2006 exhibition marked the beginning of a process whereby *Photo Op* has repeatedly been adopted, not just in its own terms, but as *the* image to complement, stand for or advertise something else having to do with the Iraq war, war in general,

Blair, or British politics. Among dozens of works featured in a major book of political art images (Klanten *et al.* 2011), it was *Photo Op* that was chosen for the cover, and the work has been reproduced numerous times in newspapers, magazines and blogs (see for example Verso 2016) and on flyers (as for example for a Stop the War Fundraiser marking 10 years since the invasion of Iraq, at the Royal Court Theatre in April 2013). Among numerous other works, it was *Photo Op* that was chosen to promote two major exhibitions of war art, *Caught in the Crossfire* at the Herbert Art Gallery and Museum in Coventry in 2013 and *Catalyst: Contemporary Art and War* at the Imperial War Museum (IWM) North in Salford (2013–2014). It also formed the cover image for the IWM book *Art from Contemporary Conflict* (Bevan 2015). At the Herbert exhibition, an enlarged version of the work was displayed on the glass frontage of the museum, facing outwards into a public square.

At the same time, while it has been commonly adopted in the public domain in relation to the war, the publicness of the work remains somewhat problematic. In 2012, Transport for London removed 100 posters featuring *Photo Op* advertising a collaborative exhibition between Hana Malallah and kennardphillipps at the Mosaic Rooms in Kensington from the London Underground, reportedly after a passenger at Green Park station lodged a complaint about the image (Mosaic Rooms 2012). In 2013, advertising companies refused to display promotional materials for the IWM's *Catalyst* exhibition that featured the work. According to the artists, one company had refused on the grounds that they did not wish to carry adverts depicting explosions on public transport (an explanation they rejected in view of the concurrent use of violent imagery to promote the new Frederick Forsyth thriller *Kill List*).[21]

Over time, *Photo Op* has also been recognised by major arts institutions, an important development in confirming the status of the work as art, but this has taken place in ways that reveal further its complexity as image, object and event. The Imperial War Museum (IWM) acquired a version of the work produced in 2007 (item Art.IWM 17541), and appear to accord the work the status of a representation of an actual event, in that they describe it in online notes as an 'image' of 'former Prime Minister Tony Blair photographing himself with a mobile phone against an exploding fireball in a desert landscape' (Imperial War Museums 2017c). In 2009, the Scottish National Portrait Gallery acquired a version of the work (item PGP 597.1), describing it as a 'Photograph' (National Galleries Scotland 2017), again implying its status as a record of an actual event. While the acquisition of an unauthorised work of a former Prime Minister by a major national gallery is notable in its own right, a further problematic aspect of the composition is reflected in the fact that no image of it is made available on the gallery website, where there is an apologetic statement that 'this image is either subject to copyright restrictions or has not yet been photographed' (National Galleries Scotland 2017). In 2010, the National Portrait Gallery in London also acquired a version of the work for its reference collection (item NPG D42643), with it being described here as a 'process printed poster' but again with no image provided online, 'for copyright reasons' (National Portrait Gallery 2017).

Photo Op has also been appropriated and adapted for other purposes, further underscoring its centrality as motif and model for political positions in Britain and beyond. In 2009, an image clearly modelled on kennardphillipps' work was used to promote a production of Bertolt Brecht's 1939 play *Mother Courage and Her Children* at the National Theatre, with lead actor Fiona Shaw posing with a cameraphone in a similar posture to Blair, in front of an explosion. In 2015, the Turkish magazine *Nokta* was suppressed after it used as a cover image a montage of President Recep Tayyip Erdoğan apparently posing for a selfie in front of a coffin being borne by Turkish soldiers and draped in a Turkish flag, a comment on a statement by Erdoğan that a soldier killed by the Kurdish PKK group would be happy to have martyred himself for such a cause.[22] Ahead of the 2015 general election, the *Times* cartoonist Peter Brookes created a version of *Photo Op* (annotated, 'with apologies to kennardphillipps') mocking Blair for warning against a vote for the Conservatives in the forthcoming general election. Brookes (2015) added two speech bubbles to the composition, the first stating 'DON'T VOTE TORY...' and the second 'IT COULD LEAD TO **CHAOS**!', with the clear implication that, in the light of the Iraq war, Blair was in no position to admonish anyone with such a claim. Later in 2015, a version of the cartoon with the text edited to refer to Blair's intervention against the candidacy of Jeremy Corbyn for leader of the Labour Party was circulated online, with the text reading, 'DON'T VOTE CORBYN... IT COULD LEAD TO **CHAOS**!' (Lee Jones 2015).[23] In response to critical coverage of Corbyn's leadership bid, the photomontage artist beaubodor (2015) reworked *Photo Op* with Corbyn in place of Blair (Figure 7.5). In order to emphasise what the artist saw as the contrast between the two and take issue with media representation of Corbyn, the words PEACE CRIMINAL were blazoned across the top, in the white capitals commonly employed in online memes.

Photo Op has also been reappropriated and adapted by kennardphillipps themselves. In January 2011, the artists projected an animated version of the work (including moving flames) onto the front of the grand Central Hall in Westminster, across from the conference centre where the Chilcot Inquiry was taking place, on the evening before Blair was interviewed for a second time (kennardphillipps 2011). In this version of the work, the artists added scrolling text above the image of Blair, reading, 'The journey stops here'.

In continuing to circulate, to reappear and to be reappropriated, *Photo Op* continues to do some kind of work in expressing the relation between Blair, the event of the Iraq war and British politics more generally, being described as 'an icon of our time' and 'the definitive work of art about the war' by Jonathan Jones (2013) of *The Guardian*. We can see the work as emerging from events, as I have emphasised, not only in relation to the Iraq war but in relation to the genealogy of photomontage itself, as well as in relation to the experimental practices of the

Figure 7.5   *Peace Criminal*, beaubodor (2015). Image courtesy of the artist.

artists. In *Photo Op*, art history and geopolitics pass through each other. In constructing an unreal, but 'true' event, and it its many appropriations and reworkings, *Photo Op* offers a counter-actualisation of the event of the Iraq war as one in which justice is served upon Blair for his role in initiating the disaster. In so doing, however, *Photo Op* both assumes knowledge of, and refers back primarily to, a British political situation (Barry 2013) and a British-centred geopolitical imaginary; that is, it 'works' in relation to the specificities of British geopolitical culture (Toal 2008). Iraq as such does not really appear. While the work operates in part through its exploitation of what Jenny Edkins (2015), following Deleuze, calls face politics, a burning oilfield stands in for the invasion and for a destroyed country. In this way, *Photo Op* works along the grain of an existing distribution of sensibility; it exploits the reductive, affective and emotive affordances of images of people and of their postures and faces, while coinciding with the rise of the selfie as a ubiquitous technological and embodied form.

Conclusions

The idea of the 2003 Iraq war as a mistake and a disaster for which Tony Blair bears primary responsibility came to be widely and deeply felt across segments of the British population. But this structure of feeling was not corroborated (Rancière 2010) by anything that could be construed as either political or legal accountability for his actions. It was in this context that a work like *Photo Op* could come to achieve the salience that it did. While many Weimar-era montages and several of Rosler's works in both series of *Bringing the War Home* are marked by considerable complexity, as are other works by Kennard and kennardphillipps, *Photo Op* is both richly ironic and devastatingly simple. As Kennard has written,

> It's often the case that, because photomontage is an additive form, one goes on adding and adding. The key is to subtract to the point where often only two main elements lock together. (Kennard 2000: 83)

This statement holds particular force with regard to *Photo Op*, which I would suggest has become widely distributed precisely because of its seeming ability to express in a single flash a widely held political truth-feeling about the war. There is just enough reality in the image for a viewer to think it might depict an actual event. But in contrast with *Know Your Enemy*, the work retains a sufficiently disjunctive quality between its two main elements such that, when closer attention is paid, it does not quite construct what, following Kriebel (2014), could be called a fully sutured reality. The composition of the work elicits a narrative about the nature of the event that appears to have been captured in it, but in so doing also calls up knowledge and feelings its viewers may already hold about Blair and the war. It only therefore functions when viewed by people who bring to it *a priori* knowledge of Blair and his relationship to the war; its affects and effects are buoyed (Rancière 2010) by pre-existing dissensus over the Iraq war, of which it provides a wordless encapsulation: phoney politician, burning oilfields, warmonger. In the absence of formal accountability, it offers a reckoning that accords with, and helps to perpetuate, a widely shared structure of feeling about the war as event.

At the same time, despite its apparent simplicity, and as I have argued, kennardphillipps' *Photo Op* bears multiple relations to the events that it counter-actualises as well as to their other works and photomontage more generally. In this chapter, I have argued for a significantly revised and expanded account of photomontage that recognises the multiplicity of the form. Such investigation needs to take account of the genealogy of photomontage as form and the complexity of its relations to events that it might be said to counter-actualise. More attention also needs to be paid to the ways in which photomontages are produced, circulated, received, reworked and redistributed, that is, to the ways in which photomontages emerge from practices and how they are integrated within and altered by them. This also suggests a move away from declarative statements about the critical or uncritical affects and effects that a work might elicit or produce. What is required, rather, is a more thorough inquiry into the multiple ways in which photomontages emerge from, constitute, and are implicated within, events, and reflection on the ways in which they might be said to map and model the virtual patterns and forces out of which those events are actualised.

During the 1990s, Kennard had to some extent turned away from photomontage because of the increasing media use of Photoshop to alter news photographs: could photomontage really continue to function if images could now be manipulated at a minute level? Contrary to expectations, however, geopolitical photomontage seems to have flourished and to have become a truly popular form, with montages emerging online almost immediately in response to unfolding events. This situation has been facilitated by the ready availability of advanced image processing capabilities on personal computers and mobile phones and by the explosion of visual social media, which are geared to immediate, affective/effective images with minimal textual content. In a context where myriad actors strive to capture, disseminate and exploit geopolitical imagery, and others rapidly examine and critique its veracity, further attention is warranted to the ways in which events are appropriated, reassembled and counter-actualised through montage practices.

Notes

1   I am grateful to Martha Rosler for several corrections, clarifications and qualifications to my reading of her work.
2   August Jordan Davis (2013) has offered a more extensive consideration of Rosler's work and her own account of it, in which the idea of the event is implicit in discussion of return, doubling and repetition; here I aim to draw out the question of the event still more explicitly.
3   As Rosler has also written: 'I was trying to show that the "here" and the "there" of our world picture, defined by our naturalized accounts as separate or even opposite, were one' (2004, p.353).
4   My thinking here is also informed by historically oriented works, including that of Patrizia McBride (2016), who argues that the elicitation of narratives is central to how photomontages work. Matthew Biro (2009) has reinterpreted Dada and post-Dada photomontages as cyborg assemblages, while Sabine Kriebel (2014) has reinterpreted the anti-Nazi photomontages of John Heartfield.
5   As Rosler has said of the Vietnam works,

> everything in these images is quite still. In fact, I was an *anti*-expressionist but it is inevitable that when we see these confrontations, especially for the first time, that the horror comes first, and that is a legitimate emotion. But I don't think that horror is my genre. I really prefer to stay quite cool, even to the point of stasis, rather than showing a lot of blood or providing some kind of moment of shock. (in Blazwick 2008, p.3, emphasis in original)

6   Individual photomontages were published in the art press sporadically during the 1990s; they became recognised as 'art' after an article in *Art America* during the Gulf War in 1991 and were then exhibited as such for the first time (Rosler 2004, p.356).

7   The relation between the visual representation of the Vietnam war and the 2003 Iraq war was explored critically in the 2008 Brighton Photobiennial, curated by Julian Stallabrass (Stallabrass 2013). In Chapter Eight, I discuss one of the works exhibited there, Thomas Hirschhorn's *Incommensurable Banner*.

8   Rosler has also stated, 'I do prefer that there be a multiplicity of cues so that ordinary people might understand where a work is going, even if they don't grasp all of its references, vectors, dimensions, and so on. There should be an element that suggests that the work is a step away from the "what is," that it operates on a meta or symbolic level.' (Courtney Fiske interview with Martha Rosler 2013)

9   Over and above their distinct US/British contexts, there are differences between the rationalities and tactics employed by Rosler, whose work is explicitly feminist in orientation and which links the problem of war to domesticity and the home, and by Kennard/kennardphillipps, whose work often links well known political actors with geopolitical and economic violence. A fuller consideration of photomontage in relation to geopolitics would explore these and other issues.

10 Kennard showed work alongside Iraqi artists as part of the 1991 *Artists Against the Gulf War* show at the Kufa Gallery in Westbourne Grove, London, now defunct but once an important hub for Iraqi diaspora culture and politics.

11 See https://www.youtube.com/watch?v=-gz88hWbcXY&t=702 at 2m40s (accessed 18 April 2018).

12 https://www.kennardphillipps.com/blood-on-the-scanner/#more-112 (accessed 18 April 2018).

13 At the opening at the Henry Peacock Gallery, outside the gallery and against a backdrop of the works reproduced as placards, Harold Pinter read poems composed in response to the invasion. See https://vimeo.com/65377031 (accessed 18 April 2018).

14 A still image of this work can be seen at https://www.kennardphillipps.com/lakenheath-american-airbasesuffolk-2005/ (accessed 9 April 2018).

15 It was reported that a 12-year-old boy had also been detained. The violence and subsequent actions of the armed forces were severely criticised by the Judge Advocate, as well as by lawyers and commentators.

16 In 2004, the editor of the *Daily Mail* had been sacked for publishing photos that were presented as recording the abuse of an Iraqi man by British troops, but which were subsequently accepted to have been faked (BBC News 2004). The faked images, however, turned out to be similar in certain respects to the actual ones later revealed during the proceedings discussed here. The contrast between kennardphillipps' art and the fallout from the *Daily Mail* scandal reveals the different kinds of truths enacted by and expected from different *dispositifs*.

17 Here I expand and develop the discussion presented in Ingram (2016).

18 This sequence has been evident to me on many occasions where the image has been shown to people individually or in groups. However, whether and how the image works depends very much on what its viewers bring to it, underscoring the importance of reception. While sometimes there is immediate laughter and occasionally people are moved to speak out spontaneously (often along the lines of, 'yes, that's it!'), some people do not know what to make of the image; some have taken it to be

of an actual scene and are embarrassed to learn that it is a fabrication. When presented outside of the UK, much more explanation of what the image is and its relationship to the event is typically necessary (although some people still 'get' it). My experience of showing the work in undergraduate classes in recent years suggests that it is also losing its affectivity and effectivity with younger people who have only an indirect knowledge of Blair and the war, or none at all about either. In these contexts, much more explanation of what it is and what it is about is typically needed.

19   This action is discussed in Routledge (2005). Photos of the group using *Photo Op* can be found on the artists' website and elsewhere online.

20   See, for example, the gallery on kennardphillipps' website, https://www.kennardphillipps. com/santas-ghetto-2006/ (accessed 17 December 2017); see also the image at Getty Images, which incorrectly describes the work as a piece by Banksy: http://www. gettyimages.co.uk/event/new-art-work-by-banksy-revealed-72678589#man-takes-a-picture-of-a-piece-by-guerrilla-artist-banksy-at-santas-picture-id72690270 (accessed 17 December 2017).

21   See: https://www.kennardphillipps.com/iwm-photo-op-advert-banned-cbs-outdoor-clearchannel-jc-deceaux/ (accessed 18 April 2018). See also BBC News (2013).

22   As reported at https://www.kennardphillipps.com/turkish-magazine-nokta-get-trouble-remake-photo-op/ (accessed 18 April 2018). As reported, the editor in chief was arrested and copies of the magazine were seized and taken off newstands. See also, https://www.theguardian.com/world/2015/sep/15/turkish-magazine-nokta-raided-and-copies-seized-for-mock-erdogan-selfie (accessed 17 December 2017).

23   This work bears the same date as the original and does not appear in Brookes' Twitter archive, suggesting it is an appropriation by someone else rather than a new work by him.

Chapter Eight
Geopolitical Bodies

This book has conceptualised the geopolitical event as a disruptive transformation of the world, of sense and sense-making, that flows through and affects bodies of diverse kinds. In this chapter I explore a series of artworks emerging out of the 2003 Iraq war in terms of how they involve human and more-than-human bodies and consider their relations to landscape, ecology and environment, tracing a distinct set of paths into, through and out of the event whose multiplicity they reveal and which they can be said to counter-actualise.

While Deleuzian thought approaches the event as an intensive flux that transforms and leaves traces upon the bodies and states of affairs through which it is actualised (Patton 1997), posthumanist thinkers (e.g. Braidotti 2013) and sympathetic critics of the term (Haraway 2016) have called for accounts of embodiment that go beyond the limits of human corporeality both to decentre the human and to track its extension via technology and the environment. Such concerns have been taken up and extended in Deborah Dixon's (2015) feminist materialist work on geopolitical bodies, which takes account of such matters as flesh, bone, abhorrence and touch, and extends out to encompass the anthropocene as event, condition and problem. Dixon seeks to develop a 'body-aware social theory' that is attentive to 'the co-constitution of corporeal bodies and the environments within which they emerge and transmogrify' (2015, p.22), and draws attention

Geopolitics and the Event: Rethinking Britain's Iraq War Through Art, First Edition. Alan Ingram.
© 2019 Royal Geographical Society (with the Institute of British Geographers).
Published 2019 by John Wiley & Sons Ltd.

not only to the complex interrelations that continually rework the corporeal geographies of the body, but also other temporalities, such as the evolutionary tempo of species and the deep history, or 'Earth history,' within which all life, as well as the materials that enable and aver this combination, is embedded. (2015, p.23)

Dixon's feminist materialism is thus sensitive to the question of 'how the planet itself has come to be configured as a geopolitical body' (2015, p.23) through Western geopolitical practices. As she further argues (also Yusoff 2015) drawing in part on Rancière, an engagement with aesthetics and aesthetic practices can help 'in the apprehension of, and living with, the Anthropocene' (2015, p.124).

A focus on the event can bring to the fore the often violent ways in which geopolitical configurations and orders are both unsettled and imposed. In this chapter, I explore the manner in which artworks might be said to appropriate and rework such processes, thereby making something else of what are often traumatic, and indeed unspeakable, occurrences and experiences. In so doing, these works also often work across ethical and political dilemmas and problems that defy attempts at resolution, themselves remaining 'problematic and problematizing' (Deleuze 2015, p.55) rather than offering solutions. If we also turn analytical attention back onto the *dispositif* of art in relation to which artworks emerge, we gain further insight into what are often complex entanglements with the events and forms of power that they might be said to counter-actualise.

The first part of the chapter discusses a single artwork, *The Incommensurable Banner* by Thomas Hirschhorn, which appropriates horrific images of injured and dead human bodies and which raises a number of ethical as well as aesthetic problems. The second considers the appearance of British military bodies in ways that reflect sympathy, concern and an effort to memorialise, and which position the human body in relation to landscape and environment. The chapter then considers artworks, projects and exhibitions emerging primarily from the work of artists and curators of Iraqi heritage that might enable us to think about geopolitical bodies in more-than-human ways, while resisting the organic metaphors of nation or state upon which, as Dixon observes, classical geopolitics has been founded. The chapter then moves beyond Britain to consider how the war (still in some ways understood as in part 'Britain's') has been addressed elsewhere in the international art world. Examining recent editions of the Iraq National Pavilion at the Venice Biennale and the critiques that have been made of it by some Iraqi artists, curators and writers, the chapter returns to the question of how, while offering a distinct sensorium that enables experimentation with events, the *dispositif* of art remains entangled with geopolitical power in myriad ways. The chapter ends by discussing the work of Sakar Sleman that was presented at the 2017 Venice Biennale, which might be interpreted as embodying some of the concerns of a decolonial, feminist and materialist geopolitics.

Injured Bodies

In October 2008, a work by the Swiss artist Thomas Hirschhorn with the title *The Incommensurable Banner* went on display at Fabrica, a former church in central Brighton. The exhibition of this work formed part of the Brighton Photobiennial, which that year sought to explore and compare the role of photography in the Vietnam war and in the wars in Iraq and Afghanistan (Stallabrass 2013). The work, its exhibition at this time and place and the biennial itself each formed an attempt to grapple with ongoing wars historically and critically. On one level, the presentation of the *Banner* at the Photobiennal can be regarded as an attempt at a direct geopolitical act that sought to bring home the brutality of modern warfare and its sanitisation by news media, but which in the process raised further problems.

The Incommensurable Banner took the literal form of a banner, some 18 metres long and 4 metres high, which, as exhibited at Fabrica, was not quite fully unfurled. The use of the banner as form already framed the images contained in the work in a polemical way: the form of the artwork signalled that it was intended to be taken as a protest. Before the viewer entered the exhibition space, however, they encountered a statement warning them about what they were about to see. In order to view the work, the viewer had to navigate around a large screen. Encountering the work required a conscious choice and embodied performance by the viewer, to enter the sensorium of art. Along the top of the banner ran rough, spray painted text, reading 'THE INCOMMENSURABLE THE INCOMMENSURABLE'. Underneath, and taking up most of the banner's space, were reproduced dozens of full-colour photographs of human bodies that had been mutilated by modern weaponry in Afghanistan, Iraq and elsewhere in the Middle East. As notes accompanying the work explained, the images derived from wars in which Western powers including Britain were involved, and which were circulating online, but which were not generally being made available to Western publics via established media outlets. The work therefore sought to stage an event by making visible pictures usually absent or excluded from public view. However, the people depicted were not identified and the circumstances of the individual events in which they had been injured and died were not described.

The Incommensurable Banner can be viewed in a long series of artworks attempting to confront war by presenting its effects on human bodies, a series that stretches back to Goya's *Disasters of War*, which depicted the horrors of the Peninsular War in the early nineteenth century, and beyond. This series would also include *Black Smoke Rising* (discussed in Chapter 6) as well as *Know Your Enemy* (Chapter 7). In implying the centrality of bodily integrity to ideas of the human and the person, the *Banner* can also be seen as raising, in an extreme way, issues that are also addressed in feminist accounts of framing and vulnerability (Butler 2004, 2010) and of horror (Cavarero 2011). As Elaine Scarry has written,

> while the central activity of war is injuring and the central goal in war is to out-injure the opponent, the fact of injuring tends to be absent from strategic and political descriptions of war... (1985, p.12)

Within the parameters of its installation at Fabrica, *The Incommensurable Banner* attempted to display the 'fact of injuring' as recorded in the photographs that it reproduced. The presentation of horrific images in a decontextualised manner, however, all but foreclosed the possibility of empathic connection with people and communities undergoing this kind of violence (Jill Bennett 2005). Lacking further contextual information, the work displayed the effects of particular kinds of events on bodies and invited reflection on the kinds of forces that might have been involved, while, by means of its title, questioning its own relationship to those events, and, by means of its form (a tool of protest located in an art exhibition), questioning an apparent lack of wider political action. In pointing up the highly mediated way in which images of conflict reached Western publics via established channels, the work also highlighted its own inadequacy as a form of representation and politics.

Military Bodies

Questions of the geopolitical body have been posed in many other ways in artworks emerging out of the 2003 Iraq war, including works that have addressed British military experience, in which soldiers have figured as artists, subjects and participants. These works can be situated in relation to a trajectory of concern for military bodies that was expressed especially strongly in British artists' responses to the First World War, in works that addressed not only the injured or dead bodies of soldiers (e.g. *Paths of Glory*, CRW Nevinson, 1919) and traumatised landscapes (e.g. *The Menin Road*, Paul Nash, 1918) but also the means of inflicting injury and death, including mustard gas (e.g. *Gassed*, John Singer Sargeant, 1919), and new military and industrial machines (e.g. *Swooping Down on a Taube*, CRW Nevinson, 1917). As Paul Gough has suggested,

> it is in th[e] interstice between revulsion and fascination that we find some of the most impressive and moving art of the Great War; paintings, prints and drawings that are a mesh of tense ambiguities, ironic counterpoint and taut poetics. (2010, p.319)

The tensions and ambiguities expressed in these works often reflected broader modernist concerns and the influence of Vorticism in particular, a movement of which several British artists who experienced and responded to the war had been part. Such concerns with the fate of human and more-than-human worlds in the face of new technologies of destruction and the violence of modernity were also central to German expressionism and to Dada, and to wide-ranging reparative efforts in Britain, Australia and the United States, in which classical ideals of bodily integrity were also important (Carden-Coyne 2009). While the scale of British military losses in Afghanistan and Iraq is many times smaller than during the First World War, reparative efforts centred on the military body have again become increasingly evident since around 2007, when a sense that the relationship between the government, military and society had broken down began to crystallise in elite circles. The perceived urgent necessity of such efforts had been intensified by the growth of vernacular (rather than officially orchestrated) practices of commemoration in the town of Wootton Bassett, through which the bodies of dead British soldiers would be transported on their return from Iraq and Afghanistan (Jenkings *et al.* 2012), and by contentious artworks including Gerald Laing's *Truth or Consequences* and Steve McQueen's *Queen and Country* (discussed in Chapter 4).

Since 2007, reparative practices have taken many different forms, including the development of new charities; the provision of more prominent roles for service personnel, who have sometimes sustained life-changing injuries, at occasions where a sense of national identity is forged, such as the annual FA Cup Final and the 2012 London Olympics; and the introduction of an annual Armed Forces Day. These are efforts in which art projects have often been called to participate, as, for example, in art therapy courses run by military charities. As much British art of the First World War shows, however, while political actors have often hoped that art would assist in the attempt to reassemble individual and collective bodies around a particular idea of the nation, and while reparative efforts may bring benefits for some people, the results of this process are frequently ambiguous, if not antagonistic, with regard to this end. It was perhaps ill advised for Prime Minister David Cameron to announce the government's plans for the commemoration of the First World War in 2017 as a 'truly national moment' while standing in front of Nash's *Menin Road*, a stunning modernist indictment of the military destruction of bodies and landscapes.[1]

The art-historical legacy of the First World War has been appropriated critically in works by British artists focusing on the harms to Iraqi bodies and landscapes, as in the engraving *Victory Palms* by Emily Johns (Figure 8.1). Much of Johns' practice has since the 1990s been concerned with the impact of sanctions and military violence on Iraq. *Victory Palms* links the palm, a symbol of peace as well as victory in Mesopotamian and Near East cultures, to the punitive destruction of date palms by coalition forces in the 2003 war, but visually the work also references Nash's paintings of traumatised landscapes. In this way, otherwise widely separated geopolitical and art-historical events are condensed within the frame of a single, painstakingly constructed image and made part of the same Event.

Figure 8.1   *Victory Palms*, Emily Johns (2008). Image courtesy of the artist.

The ambiguity of artistic representations of military bodies is evident in the drawings and paintings of Arabella Dorman, created during and following periods embedded with British units in Iraq and Afghanistan. These works in certain respects follow established traditions of British war art, in providing figurative records of the deployment of soldiers and military vehicles in overseas locations, and of local people with whom they interact, while eschewing the direct representation of violence. Like other examples of war art oriented towards military experiences and audiences, these works often evoke European Orientalist paintings of the nineteenth century in their depictions of landscapes and people. They are thus both expressive and constitutive of a particular community of sense, through which the war was enacted and experienced, but within which critical orientations towards it also coalesced, while rarely being crystallised or expressed openly. It is, rather, in depictions of exhausted, shocked or injured soldiers that one might locate a critical structure of feeling concerning Iraq as a 'bad' war.

A concern with embodied military experiences and landscapes is also evident in the work of painter Douglas Farthing, a former British soldier who served in Northern Ireland and in former Yugoslavia as well as Afghanistan and Iraq, and especially in the 2009 drawing *The Enemy Within: Driving in Iraq*

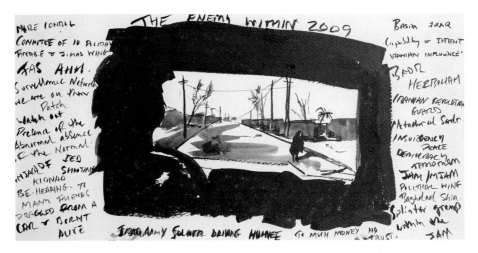

Figure 8.2   *The Enemy Within: Driving in Iraq,* Douglas Farthing (2009). Image courtesy of the artist.

(Figure 8.2).[2] This work draws upon and expresses aspects of Farthing's experience serving in Iraq, at a time when the vulnerability of British forces to attack from insurgents assisted by Iranian military and intelligence services was becoming an important political issue. The work deals with the experience of driving through southern Iraqi cities in a 'Humvee' vehicle, and is centred on the narrow view over the driver's shoulder and through the front windscreen, out onto a street, which is rendered in perspective with telegraph poles, trees and low buildings extending along each side. A small cloaked and hooded figure, perhaps female, is visible in the right foreground. The street view is delimited by a heavy black border, emphasising the constrained view of the landscape. The central pictorial scene is surrounded with roughly-composed text, including the work's title and date above, and a note underneath ('Iraq Army soldier driving Humvee'), but also fragmentary words and phrases taken from briefings in Iraq and recalled from other deployments. The sense of being watched ('surveillance network') is doubled by an over-riding need for constant scrutiny of the landscape and figures within it ('Watch out... [for the] presence of the abnormal [and the] absence of the normal').[3] The territorial context of the deployment, meanwhile, is signalled in the phrase 'we are on their patch', a phrase that could refer both to the territoriality of Iraqi insurgents and to cross-border Iranian influence (an awareness of which is further signalled down the right hand side). In the context of the work, words like 'peace', 'democracy' and 'freedom' appear abstract when opposed to the specificities of 'IED', 'shooting', 'kidnap', 'be-heading' and 'to [perhaps, also, or alternatively, 'too'] many friends dragged from a car and burnt alive'.

The human figure on the street carries particular importance, explained by the artist in ways that highlight a particular form of objectifying military visuality:

> you call them BMOs... blue moving objects or black moving objects. You can be driving along the road and looking out the window, and on the right hand side, 'You've got some BMOs' and it's that brevity of communication that's important to soldiers... the same terminology is used in Afghanistan and Iraq. It's all American, they are Americanisms which have crept into our military. But then we used the same kind of stuff in Northern Ireland because it's all about speed and being able to react quickly.[4]

The Enemy Within, then, emerges from direct military experience of the Iraq war and other conflicts in which British military forces have been involved, but it is neither glorious, nor commemorative, nor reparative. Nor does it address the politics of the war directly. Its title might refer to insurgents or Iranian agents, but could equally refer to the British forces themselves, or even to the soldier's own internal narratives; as Farthing has stated, 'My paintings tell the story of hidden tension, a hidden war'.[5] The work expresses a specific form of embodied military visuality – the highly restricted gaze from a moving vehicle out into a landscape that is assumed to be hostile – but links this with the multiple contexts and anxious narratives that give it meaning and resonance. One must also acknowledge the situated nature of the work: the viewer is invited to take the place of a British soldier and participate in his experience, and not that of the people on whose 'patch' British forces are operating.[6]

The military body has also been addressed in Kryzstof Wodiczko's *Veterans Project*, which has worked with former soldiers suffering from post-traumatic stress disorder in the US, Poland and Britain. As Wodiczko has stated, 'war veterans are a special group or a special category of 'strangers', because they come back home... but it's not a home'.[7] In the British part of the project, Wodiczko collaborated with six former soldiers contacted via the veterans' mental health charity Combat Stress (founded after the First World War as the Ex-Servicemen's Welfare Society).[8] While the charity provides art therapy courses among the services it offers to veterans, Wodiczko's project employed distinctly modernist tactics to bring veterans' experience into the public sphere in striking and decidedly critical ways.

The *Veterans Project* employed tactics developed in Wodiczko's practice over many years: participatory work with people encountering social exclusion; the construction of improvised vehicles that respond to their needs; audio and visual projections; and dramatic public space intervention. As realised in Liverpool in 2009 and presented in a 2012 exhibition at the Work gallery in London, the project drew, first, on interviews with British soldiers who had served in the Falklands, Bosnia, Afghanistan and Iraq about their wartime experiences and subsequent difficulties with PTSD. Rather than presenting these in a personalised or narrative format, selected statements and words from these interviews were extracted and assembled in a manner that, recalling the modernist cut-up technique, detached them from specific individuals and reassembled them in a new technical, sonic and visual assemblage. Having been extracted from interviews, words and statements were projected in, and into, public space from a specially constructed War Veteran Vehicle. While in the US, this was based on a Humvee, in Liverpool it employed a Land Rover, carrying a visual projector and public address system in place of a weapons platform, a transformation that, in Wodiczko's (2011) telling, turns a 'projection battle station' into a 'projector battle station', able to bring the war into the public sphere back home in an effort to describe what he calls 'the wall' that separates those who have been transformed by war from those who have not.

The vehicles are deployed in public spaces at night time, when rapidly projected words are visible on the sides of buildings and the soldier's voices, interspersed with the shocking sound of loud, rapid-fire gunshots, fill and reverberate around the surrounding area. While soldiers do not appear in corporeal form, the vehicles form a temporary, evental assemblage between their voices, their words, appropriated military technology and urban space, enacting temporary publics or, in Rancière's terms, a community of sense.[9] This community is composed out of a jarring conjunction of sensation and sense making: as one veteran interviewed by the BBC stated, 'You've got to be in these situations to understand the emotional side and the effect it has on you' (BBC World Service 2009).

The vehicles are intended to be, at the same time, useful, carrying veteran's neglected voices and critiques of their experiences physically into the public domain, and disruptive of official propaganda surrounding military recruitment and war, while functioning as monuments to how veterans are transformed into different people by their recruitment and their wartime experiences. Wodiczko also describes the importance of the urban settings, in ways that further point towards an extended and more-than-human conception of the body:

> Public buildings are witnesses to many events that took place in our cities. They remember all sorts of things. But it's important to also understand that soldiers are silent monuments to their own trauma as well. So, here, if they can animate the silent, symbolic architectural structures and teach them how to speak, then they will convey their voice with greater strength and magnitude than speaking directly. At the same time, in order to animate those structures they have to animate themselves. They have to develop their capacities to open up and speak. (BBC World Service 2009)

The tactics evident in the *War Veteran Vehicle* project connect in certain respects with Rancière's idea that 'politics revolves around what is seen and what can be said about it, around who has the ability to see and the talent to speak, around the properties of spaces and the possibilities of time' (2006, p.13). The vehicles construct a community of sense in which veterans' voices are given a chance to resonate discordantly in public space, disrupting its 'primary aesthetics' (Rancière 2006, p.13), bringing the war home in a different way. The project, as realised in Liverpool in September 2009, was also expressive of a structure of feeling discussed in this chapter and elsewhere in the book, in which widespread disquiet at the war, stemming from many different points, was in the process of coalescing and finding aesthetic-political expression.

The framing of the project bears a relation to a further dimension of the distribution of sensibility. As Wodiczko related,

> There is a wall between those who know what war is and those who don't. It's a very thick wall [and] [t]he only people in a position to try to break this wall are ex-soldiers themselves... (BBC World Service 2009)

One cannot read too much into Wodiczko's words here: soldiers undoubtedly face difficulties in crossing back from the military sphere to the civilian following war, and he might have been expressing the critical point that it is only soldiers' voices, among the many others who might speak, that have a chance of being heard. But it is nevertheless reflective of a distribution of sensibility in which, especially, people of Iraqi heritage, their voices and their aesthetic and political work, are often also either invisible and inaudible, or are present in a way that is conditioned by Orientalist stereotypes, or are in some way secondary and subsidiary to an assumed dominant majority, and which thereby imposes relations of inequality. This is not to diminish the problems faced by military personnel. Rather it is to point up how hard it is for Iraq and Iraqi people to appear even in critical public accounts of the war in Britain. In the remainder of the chapter, I focus on a series of projects by, and involving, artists of Iraqi heritage that have been realised in Britain and beyond, in which the transformation of geopolitical bodies in the course of the event is at stake, but which also reveal the complex of implication of the *dispositif* of art in geopolitical power, and how this, too, alters in the event.

More-Than-Human Bodies

Deleuzian thinking approaches the event as an intensive incorporeal flux that transforms and leaves traces upon the bodies and states of affairs through which it is actualised. As well as informing feminist geopolitics, this idea has also been developed to consider the role of events in the formation of 'bodies politic' (Protevi 2013). Such thinking poses particular ethical challenges for thinking through wounding and death, not just as events in themselves, but as models for the event as such (also Scarry 1985; Cavarero 2011). Deleuze's (2015) work further explores how the violence of events and the traces they leave may be confronted, transmuted, disengaged and 'differenciated' through acts of will and creativity, that is, through the kinds of counter-actualisation that art enacts. These ideas can be linked with arguments about colonialism as a wound in decolonial thought (Mignolo 2010; Vazquez and Mignolo 2013) and the need to attend to and confront the bodily harms of war in feminist political theory (e.g. Butler 2004, 2010) and feminist political geography (Hyndman 2007). Advocates of posthumanism (e.g. Braidotti 2013) and its sympathetic critics (Haraway 2016) have also called for accounts of embodiment that decentre and go beyond the limits of human corporeality.[10]

While my approach here does not centre on 'the matters of sex, sexuality and reproduction' that are of particular concern to Dixon (2015, p.19; see also Grosz 2011), it shares her broader interest in 'differential orderings of and access to life on Earth', as well as the specific focus on 'differential renderings of corporeal vulnerability and obduracy' and an interest in how such concerns are both explored and forwarded through art. The discussion here is driven by a series of projects by and involving artists and curators of Iraqi heritage that have sought to reframe and reassemble more-than-human geopolitical bodies out of, against and, to some extent, beyond, specific experiences of colonialism and war amidst a growing recognition of the predicament of the anthropocene (Davis and Turpin 2015). In taking us well beyond communities of sense oriented around the British imperial and colonial nation, state and military, this discussion also returns to threads developed earlier in the book in relation to the geopolitics of museums and oil. This shift in perspective also alters the sense we might have of the Iraq war as ongoing geopolitical event. An important aspect of the artistic and curatorial work I discuss here is a structure of feeling within which it is felt important to rework and move away from the thematics and images of abuse, destruction and trauma that have often preoccupied Western artists and curators, and to pay greater attention to how human and more-than-human lives might yet be composed within and against settings that are deeply marked by multiple forms of violence. The event is thus at stake in quite different ways.

Re-Piano, Geo-Piano

Such concerns have informed the artistic, research and curatorial practice of Rashad Selim since 2008. One strand of this work centres on Iraq's waterways, their role in shaping civilisations in the region, and the forms of material culture and practice that have sprung up around them.[11] Another strand, on which I focus here, is expressed in the *Re-Piano* project, which was prompted by Selim's experience on visiting Iraq after the invasion in 2003, as well as by multiple photographs appearing in the Western media of pianos in ruined buildings in Iraq, sometimes being played by American soldiers, and by the sight of an abandoned piano near his London studio (ArtRabbit 2010; Re-Piano Project 2010).[12] As Selim relates, the piano project 'had to do with reclaiming ... [A] lot of what I am trying to do has to do with searching for alternative narratives to things' (interview with author). As he further describes,

> [In Iraq] I had visited a music and ballet school, which my father had helped to set up. Amongst the looting I saw a lot of pianos that were destroyed, also images that I've used in other works of looters and also of soldiers playing on broken pianos in the palaces. So the whole thing with the broken pianos stuck in my head. I found a piano here [in London], dumped, with its action and keyboard ripped out, and what struck me was the metaphor of civilisation and its destruction, and its [many] relations. Also what struck me about the piano was its cover, its veil, if you wish. Without its action it is first of all a Sumerian *kithara*, the original harp and origin of the word for guitar. It also reminded me of the *santur* or *kanoon*, which is a hammered instrument, like a zither with a hammer. (interview with author)

Evoking and connecting with many other objects, the piano not only elicits multiple cross-cultural and cross-national narratives but can be interpreted as constituting a virtual event in its own right, of which individual pianos are actualisations. As Selim also relates,

> The piano is a child of the industrial revolution; before the industrial revolution you don't have this instrument popularised everywhere. [It's] because of the foundries that you could have this iron harp. And of course, because of colonialism you had the ivory, the mahogany, various woods and stuff. (interview with author)

The piano is a global, more-than-human object that brings together many different events, and which in itself constitutes a complex, ongoing event. For Selim, the piano also elicited corporeal associations and resonances, and invited experimentation: 'I engaged with the piano as sort of a devastated body, and there are all these sounds. I was quite moved by all these different sounds and started to play around with it' (interview with author). As appropriated in this project, any individual piano has the capacity to be mobilised as an evental assemblage and to counter-actualise geopolitical events.

Beginning with a single found object, the *Re-Piano* project has appropriated a series of pianos donated by people in Britain and transformed them in a variety of ways, with subsequent events being recorded in photographic and video form and incorporated in further artworks. The piano project thereby engages in its own remaking of the world and of sensation and of sense-making.

The *Atlas (Conversation) Piano* (Figure 8.3) collages the instrument with several books and photos. On the music stands rests a book with its pages opened to show photos of 'Asia', in the middle of which has been mounted a photo of a military drone aircraft. Underneath are two photos of American soldiers sitting at pianos amid damaged surroundings. Positioned under the keyboard and above the pedals is a photo of a drone control centre. The *Geo-Piano* (Figure 8.4), the

Figure 8.3   *Atlas (Conversation) Piano*, Rashad Selim (2009). Image courtesy of the artist.

fourth piano in the series, meanwhile, had its strings tuned to five scales taken from different cultures, and was extended by the addition of devices enabling it to be played in different ways. The upper surface of this piano is also collaged with a mosaic of fragments from an atlas (Figure 8.5), linking the instrument with mountains and rivers, and thereby further associating it with geographies other than those of the nation-state. As Selim describes, 'I started thinking about geography and the body and collaged into it an atlas. Having it sort of inside and on the body, but I also had the strings tuned so it *became* a geography… the piano becomes a metaphor, a theatre, a place, it becomes a geography' (interview with author).[13]

The *Geo-Piano* has been installed and played alongside photographic images, including prints of Selim's mother on an island in the Tigris and swimming in the river, taken by Selim's father during the late 1950s, a time remembered as being of particular prosperity. These images are juxtaposed with Ministry of Information propaganda slides that were prepared for a conference of non-aligned countries that never took place due to the Iran war. Though legible as a kind of critique, the

Figure 8.4   *Atlas (Conversation) Piano*, rear view, Rashad Selim (2009). Image courtesy of the artist.

work also operates as a form of regeneration, assembling temporary communities of sense through objects, devices, representations, performances and memory.

The Iraq National Pavilion at the Venice Biennale

Questions of geopolitical bodies and events have also been explored since 2011 at the Iraq National Pavilion at the Venice Biennale, which has sought to bring the work of artists based in Iraq and the diaspora into the international art world via a series of curatorial projects. Considering the Pavilion and the debates that it has occasioned offers a way to broaden the focus of discussion beyond Britain, while not sidelining its importance as a central actor in the 2003 war. Rather than providing a full account of each Pavilion, I highlight selected points of particular relevance to the arguments of the book.[14]

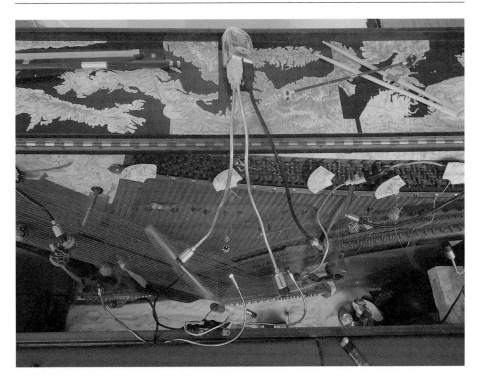

Figure 8.5   *Geo-Piano*, view from above, Rashad Selim (2009–2010). Image courtesy of the artist.

The international art circuit of biennials, triennials and so on has been extensively critiqued (e.g. Filipovic *et al.* 2010). As well as reflecting and contributing to the institutional dominance of the nation-state in its material and discursive organisation, the Venice Biennale, as the single most prestigious international art world event, is thoroughly implicated in the forms of patronage and commodification that structure the art world more generally, while also acting as a catalyst for them. Nevertheless, the Biennale continues to offer a setting in which a kind of aesthetic equality can be asserted, and in which the forms of nation and state that structure it can be opened to reflexive exploration and experimentation. The Biennale both expresses, and provides a forum for experiments with geopolitical orders, practices and events.

The staging of an Iraq Pavilion has raised several issues in terms of how its patronage, curatorial framing, thematic focus and artistic representation relate to the politics of Iraqi state formation and the diaspora (see for example Heddaya 2015). The 2011 Pavilion involved the work of six male artists, all from the diaspora, and was framed by the idea that art itself was impossible under the Baathist regime of Saddam Hussein, an expression of a discourse that delegitimises artists who remained in Iraq and the work they created there in view of their imputed complicity with the regime (on which see Davis 2005; Shabout 2007; Al-Ali 2011). While this edition of the Pavilion was important in bringing Iraq back to the Biennale, all of the artists represented were middle-aged men from the diaspora, and there was a certain eclecticism to the exhibition, with two artists affiliated with prestigious London-based galleries (Halim al Karim with Saatchi, and Ahmed Al-Soudani with Haunch of Venison) and one (Ali Assaf) who served as both commissioner of the Pavilion and artist represented within it.

While the curatorial vision for the Pavilion sought to avoid explicit reference to war and to Iraq's political history, its title – *Acqua Ferita/Wounded Water* – called up the idea of corporeal injury and alluded to an issue of immense political significance for the country and the region, while making a link between Iraq and Venice (Gervasuti Foundation 2011). Some of the works addressed the question of water directly, notably two installations by London-based artist Walid Siti. *Meso* was a rendering of a stretch of the Great Zab (or Zei) River that runs through a region that formed a centre of Kurdish resistance to Saddam's rule. The installation was made up of strips of red twill tape hung using fishing line in front of a mylar mirror and cut to trace out the winding path of the river horizontally across the work, revealing the mirror behind. Another work, a video installation, shows a famous waterfall in Iraqi Kurdistan pouring through an image of an Iraqi banknote. While appearing in a national pavilion, these artworks thus might have been taken to embody a critical orientation to the recognised geographical configuration of the Iraqi state.

Since 2013, the Pavilion has been organised under the auspices of the Ruya Foundation for Contemporary Culture in Iraq, achieving widespread international coverage and critical appreciation, and drawing in the participation of high-profile artists, writers and curators as well as artists resident in Iraq. Ruya's role has at the same time been critiqued by artists and critics of Iraqi heritage in the diaspora, in ways that bear on the relations between art, geopolitics and events. To understand these critiques, it is necessary to know that the director of Ruya, the historian, journalist and curator Tamara Chalabi, is also the daughter of Ahmed Chalabi, an Iraq-born politician who spent much of his life in exile in Britain and the US and who became one of the leading lobbyists for the 2003 invasion in Washington DC. Chalabi directed the CIA-backed Iraqi National Congress (INC), an organisation envisaged as a government-in-waiting that would take office following regime change in the country, but which was sidelined by the US government amid suspicions that Chalabi was also working with Iran.[15] Having grown up in exile in in London, Beirut and Istanbul, Tamara Chalabi returned to Iraq in April 2003 as part of a party reportedly including her father, hundreds of members of the INC, and US special forces troops (Giovanni 2010). After having worked on other charitable and philanthropic projects, Chalabi and two colleagues founded Ruya in 2012, and the organisation has gone on to coordinate three subsequent Iraq Pavilions.

The Ruya pavilions have encountered a number of public criticisms voiced by artists, curators and intellectuals of Iraqi heritage in the diaspora. While the practice of working with international curators and artists to produce national pavilions is by no means uncommon, and while argued by Ruya to be necessary in view of Iraq's long-term isolation from the international art world and suffering under sanctions and war, the role of international curators in 2013 and 2015 has been questioned (Heddaya 2015). Specific criticism has also been made of the association between Ruya and the role of Ahmad Chalabi (whose support is acknowledged in the catalogues for the 2013 and 2015 pavilions) and the INC in the war. As Dia Al-Azzawi (who was represented in the 1976 pavilion) has stated,

I find what's happening with this foundation [to be] a cover-up of the destruction of Iraq, which went from a society that is in many ways very normal to one that is a sectarian society, with a lot of killing and ethnic cleansing in different parts of Iraq. (in Heddaya 2015)

For Rijin Sahakian (2016), a critic and curator who was involved in the 2011 Pavilion and who founded a charitable project aimed at supporting young artists in Iraq that operated from 2010 until its suspension in 2016, the Ruya-run pavilions were part of a situation where 'Those who are responsible for and profit from a country's undoing also sponsor, applaud and exhibit works produced from the wretched conditions they themselves have helped sow'. Jonathan Watkins, the distinguished British curator of the 2013 Pavilion, which focused on the work of artists living in Iraq, has described being transported around the country in armoured vehicles and with armed guards, stating, 'It would be difficult to do the project without what the Chalabis can offer in terms of security and contacts' (quoted in Fernand 2013, p.44). While focusing her critiques on 'politicians' and their 'beneficiaries', Sahakian (2016) also noted that their 'endeavours are supported by an array of curators, arts writers, and financial and public relations systems that comprise the arts industrial complex'. The poet and academic Sinan Antoon, meanwhile, critiqued the focus of the 2015 Pavilion on art in light of the cultural destruction then being wrought by the Islamic State (IS or ISIS). As he wrote,

ISIS is horrible, but any reasonable person would say were it not for the US invasion there would be no such thing as ISIS... And Tamara Chalabi herself was working with her father, these are the people who brought us the dismantling and destruction of the Iraqi state. (Antoon in Heddaya 2015)

For Sahakian (2016), a key reason for suspending the work of Sada was that it was no longer possible for the organisation to continue without cooperating 'with members of the government and their supporters'; she further stated that this 'would be anathema to our mission considering the reality of corruption and context of warmongering'.

To place these sharp polemics in broader context, it should be noted that the Iraq Pavilions since 2013 are particular manifestations of a *dispositif* of art in which, while certain things are made visible and foregrounded, others are necessarily backgrounded or occluded in the process of giving art a certain autonomy, something that applies no less to military or 'embedded' artists or the British museum institutions discussed elsewhere in the chapter and book than to the Iraq National Pavilion.

In describing his approach to the 2013 Pavilion, Watkins explained his avoidance of an explicitly political framing:

> I wasn't trying to be overtly political... It's not about making work about checkpoints and bombs. Nothing I have selected sets to wring out emotion. Much of what I am interested in is about artists making do and getting by, about ingenuity. (in Fernand 2013, p.45)

Similarly, the Foreword to the catalogue for *Welcome to Iraq* stated,

> What are the first words that come to mind when Iraq is mentioned? War, destruction, displacement? Not art, creativity and resilience? The title of our show is intended to entice the audience to come and take a peek inside the country, and while it would seem that they may be met with the predictability of destruction and chaos that is present day Iraq, what they will be faced with is something quite different.

> Despite the ravages of war, Iraq's artists have still managed to find their voices and push the boundaries of their craft while dodging daily explosions, power cuts and a poor supply of artists' materials. (Ruya Foundation 2013, p.8)

Taking note of the critiques described above does not mean that the content of the Pavilions should be overlooked or discounted. On the contrary, the questions of how artists make do and get by in Iraq and how they work through issues that are indirectly related to the war are important and require attention and reflection. Indeed, it is precisely towards such questions that many artists and curators of Iraqi heritage have called for attention to be paid.

My approach is to take seriously the framing and content of the Pavilions, while at the same time recognising that the *dispositif* through which the autonomy of art is produced is also always in some way geopolitical. While this problem may be posed in particularly sharp and difficult ways in relation to Iraq, and while artists, curators and activists of Iraqi heritage have taken different positions on the Pavilion, it is not unique in this regard: as critical, feminist and decolonial histories of art have shown, the problem is pertinent to the *dispositif* of art itself at the most general level. To loop back briefly to issues considered in Chapter 6, artists, curators, critics and activists in Britain have taken different positions on the issue of oil company sponsorship of cultural institutions including the Tate and the British Museum (Miller 2015; Ingram 2017), while in her work *Is the Museum a Battlefield*, the artist Hito Steyerl (2013) has highlighted myriad ongoing connections between arms companies and Western art institutions. As I have also argued (in part inspired by Adorno 1974, 2013), artworks can be interpreted as critiquing the institutional and discursive conditions that allow them to exist as such. According to this argument, and in line with my critique of some work on art in international relations and political geography, an account of artworks that does not also consider the institutional and discursive contexts that facilitate their existence is somehow lacking. Having discussed some of the political debates and critiques surrounding the Iraq Pavilions, I now therefore also consider the framing and content of the 2013 and 2017 exhibitions, which are indeed worthy of discussion and which bear particularly directly on the questions of living through and beyond the event, its multiplicity, and the possibility of making something else of it.

The 2013 Pavilion exhibited works in a range of forms and styles, including politically-oriented cartoons (Abdul Raheem Yassir), landscape paintings (Bassim Al-Shakir), abstract painted works layered with symbols and text (Kadhim Nwir), sculpture (Furat al Jamil) and photography (Jamal Penjweny), but what I would particularly like to highlight are works that operate at the intersections between everyday life, useful objects and wider ecological predicaments.

Notable here are the cardboard sculptures by WAMI (Hasheem Taeeh and Yassen WAMI) of domestic furniture, art works and devices (such as a clock and radio), which were used to furnish rooms within the waterside palazzo hosting the Pavilion. Contrasting with the palazzo's interior, the installation created a living environment defined by a material associated with mobility, transiency and improvisation and which is readily recycled, while a sculpted suitcase suggested preparations for departure. Some of the artworks, meticulously and skilfully fabricated using differently textured and shaped pieces of cardboard, took the form of the kinds of objects that are often relegated to the status of primitive artefacts in colonial museum collections, perhaps alluding to Iraq's marginalised location within global power relations. Sculptures by Akeel Khreef, meanwhile, appropriated components from discarded bicycles and generators to create useable furniture. These works, reflective of the artist's practice working with waste materials, again contrast with their surroundings, while referencing the problem of irregular and uneven access to energy and mobility in many Iraqi cities, despite the country's massive petroleum reserves and oil and gas industry. In painting colourful motifs and text onto everyday objects like a small television, an oil heater and canteen, Cheeman Ismaeel appropriated and gave personal meaning to appliances that connect a household with much wider networks of communication, circulation and consumption. Collectively, these works point towards improvisation and endurance in the midst of disruptive geopolitical events, exploring how human lives are sustained as well as constrained via more-than-human infrastructures and materials.

The 2017 Pavilion, *Archaic*, was also organised by Ruya and involved curator Paolo Colombo (of the Istanbul Museum of Modern Art) and Belgian-born, Mexican-based artist Francis Alÿs alongside artists of Iraqi heritage from Iraq and the diaspora. This edition of the Pavilion can be read as an exploration of the emergence of Iraq as a particular kind of geopolitical body through a series of formative events and involving multiple forms of materiality, whose temporalities and spatialities both interface with, and exceed, the framework of the current Iraqi nation-state (Alÿs *et al.* 2017). *Archaic* can also be said to have engaged critically with the problem of the museum as a colonial and national institution amidst geopolitical events and power relations, going beyond the scope of previous Iraq Pavilions to present a deep and wide-ranging engagement with some of the fundamental issues and questions in the formation of Iraq as a modern state and nation, which it serialises in new ways.

The exhibition was set out in another waterside location, the former library of the Palazzo Cavalli-Franchetti, and was centred on a series of 11 vitrines, thus employing a museological (and, as Chapter 4 argues, geopolitical as well as scientific and artistic) form of display. Via the assembly of carefully selected and constructed objects, texts and materials, each of the vitrines referred to and reframed formative events in the making, unmaking and remaking of Iraq. The Pavilion can thus readily be interpreted in an evental fashion, as an evental assemblage of evental assemblages: itself an event that involves, is about, and which reserialises, many other events.

The first vitrine presented artefacts borrowed from the Iraq National Museum dating from the Neolithic (6500 BC) to the Parthian (to 150 BC) periods, establishing links with early human civilisations whose art inspired many European modernists (Bahrani 2017), and with the Museum as a key institution in the formation of the modern Iraqi state and nation. Some of these artefacts had been recovered from Syria, the US and the Netherlands, where they had ended up as a result of illicit export and sale following the 2003 invasion. The next vitrines encountered by the visitor then presented works by Jawad Selim and Shaker Hassan Al-Said, who were, as discussed in Chapter 5, formative figures in the emergence of modern Iraqi art in the mid-twentieth century, who had studied the collection of the National Museum as well as European modernism, and who went on to influence many artists working within and beyond Iraq, including Hana Malallah and Rashad Selim.

The next six vitrines presented the work of contemporary artists of Iraqi heritage exploring a variety of problems, themes and events. These included a work by Sherko Abbas dealing with the experience of his sister, the cellist Khabat Abbas, following the 2003 invasion. Khabat had been invited to join the Iraq National Symphony Orchestra on a trip organised by the US State Department to Washington DC in November 2003, during which they gave a single concert in front of an audience including George W. Bush, Condoleeza Rice, Donald Rumsfeld, and Colin Powell, who also delivered a speech on peace and democracy (Szyłak 2017). Whereas Rashad Selim's *Geo-Piano* project, discussed above, has sought to deterritorialise and reappropriate the piano, this work documented the appropriation of classical music for patently propagandistic aims as well as what appears to be the uneasy, yet professional, participation of the musicians involved. The vitrine presents video (shot by Khabat Abbas during the trip) and photographs, along with specially issued travel documentation and other visual and textual materials. One section of video shows a few seconds of footage of the flight, aboard a military transport plane, taking the orchestra out of Baghdad on the way to the US. In this, the American servicemen operating the flight and the musicians they are transporting are visible. When I encountered this work, the contrast between the grinning expressions of the airmen and the more neutral looks on the faces of the musicians intensified a creeping sense of horror at the exploitative nature of the exercise. A more recently filmed video of Khabat playing the cello in her apartment in Berlin, where she now lives, introduced a further sense of temporal and spatial distance between the 2003 event and the present, as well as lending an ambiguous reflexivity to the work.

The last section of the exhibition presented the contribution of Alÿs, which comprised two elements. The first of these was a video, displayed on a screen mounted on the library wall. The video showed a small canvas being held up by the artist, while he experiments rapidly with successive blobs of paint of different colours, apparently trying to compose a painting of Kurdish Peshmerga soldiers, two tanks and a tent in the near distance, in front of a sandy berm that is apparently providing them with cover from things happening beyond. Suddenly there is a sound like the firing of artillery, and the artist's hand roughly wipes the canvas clear, before the video restarts.

The video derives from a visit the artist made to the frontlines of the battle to retake the city of Mosul from ISIS (Alÿs 2017). In notes taken during the visit, Alÿs reflects on how each side was reshaping the landscape for tactical reasons, but also in the service of geopolitical imaginaries:

Earthworks. The Peshmerga offensive is a massive engineering enterprise, a monumental land art operation. Behind each platoon there is a bulldozer waiting. Every hundred metres of gained territory results in hundreds of tons of dry earth pushed forward, all in order to move the front line ever closer to the suburbs of Mosul. Landscape is refashioned daily by the shelling, ISIS's tunnels are behind, under, and beyond our mobile front line; the dunes are scarred by the infinite lines of trenches while on the Syrian-Iraqi border ISIS's bulldozers breach a passage through a hill to erase the Sykes-Picot Agreement's fatal design. (Alÿs 2017, p.49)

Here Alÿs appears to juxtapose ongoing events with the performative and highly mediated way in which ISIS fighters declared an end to borders stemming from the Sykes-Picot Agreement of 1916. This agreement was brought into the exhibition in the second element of his contribution, a reproduction of the notorious annotated Sykes-Picot map displayed in the final vitrine. On a map originally published by the Royal Geographical Society in 1910, Mark Sykes and François Georges-Picot sketched out a demarcation between potential British and French spheres of interest in the region, with a line running from just north of Acre on the Mediterranean coast in the west to peak in the mountains north east of Kirkuk in the east. Though conceived as a short-term fix, and while partially revised after British forces occupied the oil-rich Mosul, a section of the line went on to form part of the Iraq-Syria border. While the undue significance often given to the map has been critiqued by historians (Reidar Visser 2013; Barr 2016), its inclusion in the exhibition both reflected and extended its appropriation in struggles surrounding Iraq today.

As Sinan Antoon and other critics (e.g. Chulov 2014; Hanieh 2015) have observed, the 2003 invasion and subsequent occupation of Iraq were decisive in creating the kind of environment in which a movement like IS could emerge and thrive, and it was in order to counter IS that British military forces returned to Iraq in significant numbers alongside forces from the US and other countries. In these circumstances, the work of painter and installation artist Sakar Sleman exhibited in the 2017 pavilion suggests itself forcefully in terms of its enactment of multiple kinds of bodies, and in terms of how it has been brought into the international art world at the Biennale. As described by Kalliopi Minioudaki (2017), Sleman overcame opposition from her family, as well as poverty and isolation in pursuing her training at the College of Fine Arts in Sulaymaniyah. Since 2008, her practice has addressed a series of questions around gender, war and power, in works that engage with her own experiences of pregnancy and childbirth, the abuse of women at the Amna Suraka (Red Jail) security headquarters under Saddam (discussed in Chapter 6), and the role of oil in the region's politics. More recently, Sleman has worked with refugees from the Syrian conflict and women from Yazidi communities systematically targeted by IS. In several of these works, circles have been employed as 'structural devices' (Minioudaki 2017, p.77), as evocations of corporeal forms and as supports for memorial practices.

These concerns are expressed in Sleman's contribution to the Pavilion, which drew from her 2014 work *(Untitled) Land Art*. This piece reflects what is described as the artist's feminist humanism (Minioudaki 2017), but which engages in the constitution of earthly, more-than-human assemblages. I close the chapter with a brief discussion of these works. *(Untitled) Land Art* was realised on the side of a mountain above Sulaymaniyah, and took the form of a white circular symbol, several metres across, which was carved into the soil and marked out with stone and ground chalk with the help of several other people (Ruya Foundation 2017). Visible from a distance, the work drew the interest of people driving up the mountain, but over time was washed away by rain. Though the work echoes the land art movement and in particular the circular forms employed by British land artist Richard Long (Minioudaki 2017), it was conceived before the artist had learned about them, and while embodying the movement's original phenomenological concerns, it also moves well beyond them. As Sleman has stated,

Moving away from the studio and out onto the mountain road allowed me to produce work on a larger scale ... and to use these earth materials that children play with. ... When I touch the soil, stones and dust, I feel I am going back to my origins in nature. (Sleman in Minioudaki 2017, p.78)

There is, however, no sense that this return heals trauma or by itself achieves personal or collective reconciliation; rather it functions, in the artist's words, as an 'anti-memorial' and 'a wish for local and global peace', something it enacts by '[d]ancing with the earth as another planet' (Sleman cited in Minioudaki 2017, p.78).

In the Iraq Pavilion, the work was realised in the form of a miniature installation in one of the exhibition's vitrines, with a pile of soil and stones taken from the mountains around Sulaymaniyah bearing a rendition of the white symbol. The installation thus echoed many of the other objects on display in the Pavilion – the archaeological artefacts in the first vitrine; the palettes employed by Selim and Hassan; and the sandy soil depicted in the video by Alÿs – while questioning the cartographic political imaginary expressed in the Sykes-Picot map. It thereby served as both double and extension of the original environmental work, but one that also took on new resonances and juxtapositions as part of the exhibition. In the face of multiple violent events, it offered a counter-actualisation of Iraq as a shifting assemblage of geopolitical bodies that work through and beyond the nation-state.

Conclusion

While acknowledging the ways in which important questions surrounding Iraq's becoming as a geopolitical body have been explored through the Pavilion, it is also necessary to recognise the critical issues that have been raised regarding its own complicated and contested implications in geopolitical events and geopolitical power, the more so in light of how the Pavilion has become increasingly widely recognised across the international art world, frequently being singled out since 2011 as a highlight of the Biennale. The Pavilion has also become more firmly embedded within the international art world, through collaborations with star curators and artists, but also in the case of the 2017 project, via a series of essays from distinguished curators and authors, including writers cited elsewhere in this book. The Pavilion has become a pre-eminent site for ongoing artistic experimentation in relation to the country, but one that is, no less than any other, transected by multiple power relations, networks, hierarchies and exclusions. While some people of Iraqi heritage have chosen not to engage with, or to disengage from, those networks, they have come to form a crucial, if not obligatory, passage point (Latour 1988) for the artistic enactment of the country.

This chapter has considered how diverse artists and curators have assembled different kinds of geopolitical bodies, bearing different kinds of relations to the

Iraq war. This has underscored not only the ongoing centrality of ideas of the body as a concern for artistic, curatorial and analytical practices, but also the ways in which considering the war through ideas of the body adds further layers to our understanding of its multiplicity as a geopolitical event. Many of the interventions discussed here chime with feminist accounts of the differentiated ways in which human bodies are implicated in geopolitics according to gender, race and class, as well as other axes of subjectification, but also with the argument that, while recognising how power relations are enacted by and through individual and collective human bodies, geopolitical bodies also need to be considered in terms that are both more and, perhaps, less (Philo 2017) than human. A consideration of art emerging from the 2003 Iraq war and the disastrous circumstances and events it fostered also throws light on the continuing coloniality of the present, as well as how artists and curators have sought to engage and, in some cases, move beyond it.

A central argument of this chapter and the book as whole is that it is through the constitution of zones of provisional, partial and privileged – and therefore, invariably problematic – autonomy from geopolitical events that art offers ways of expressing, experimenting with and encountering them. While the possibilities for artworks to cross over and influence the course of geopolitical events, though not negligible, are small, and while art is constantly being mobilised towards a variety of partisan ends, we must also recognise how, in the aesthetic *dispositif*, art can always be interpreted as resisting and exceeding whatever particular political sense or purpose might be made of it.

Notes

1   Footage of Cameron's speech can be viewed at http://www.bbc.co.uk/news/uk-19913000 (accessed 19 April 2018).
2   In 2013 this work was shown as part of the exhibition *Geographies of War: Iraq Revisited* at University College London (Ingram 2013; see image at http://responsestoiraq.files. wordpress.com/2013/04/geographies-of-war-iraq-revisited-catalogue.pdf, accessed 9 April 2018). In early 2018, it was accepted into the collection of the Imperial War Museum London. The discussion here draws on the author's interview with the artist.
3   This experience is not limited to Iraq; as Farthing recalls, 'There's a terminology and again this goes all the way back to Northern Ireland, the presence of the abnormal and the absence of the normal.' Interview with author.
4   Interview with author.
5   Notes for *Comfort Zone* exhibition and workshop, Norwich 2012. http://www. douglasfarthingart.co.uk/images/Comfort_Zone_Exhibition_Workshop.pdf (accessed 1 November 2017).
6   As accounts of the British presence in Iraq have described (e.g. Steele 2009; Fairweather 2011; Ledwidge 2011), Basra was 'taken' by British troops early in the war, but they failed to establish effective control or governance of the city, or to achieve meaningful reconstruction, while antagonising and alienating its residents.

7   While the US project worked with homeless veterans, Wodiczko appears to suggest that this is also an existential condition for all returning veterans.

8   https://www.combatstress.org.uk; also http://news.bbc.co.uk/1/hi/entertainment/arts_and_culture/8272634.stm (both accessed 18 April 2018).

9   One of the vehicles was deployed close to the Democratic Party Convention in Denver in 2008; the Liverpool event took place in 2009.

10 While Jane Bennett (2010) argues that Rancière's focus on human bodies, speech and action limits his aesthetic and political imaginary, Deborah Dixon (2009) uses his ideas in conjunction with a consideration of art to rethink the human and consider more-than-human politics and aesthetics.

11 See description at http://culturunners.com/projects/art-re-imagined (accessed 7 November 2018).

12 There is a record of the project at https://www.facebook.com/repiano/ (accessed 15 December 2017).

13 Here Selim describes the piano in ways that highlight an indeterminacy between immanent and transcendent understandings of space. Selim also likens the devices added to the GeoPiano to prosthetics, while describing the deconstruction of another piano as being like opening up its anatomy. A further piano that was beyond repair was burnt, as the artist describes, 'on the lawn of an Englishman' (interview with author). Selim also describes the restorative effects of the piano project amidst the depression and despair engendered by the war and by the dominance of images and narratives of violence and abuse: 'It's got alternatives, all these different things, in it. It made me really happy after a period of feeling lost. I was doing sculpture and all these other things, but not feeling that I was engaging except in a symbolic way. It was the first thing that I felt was actually engaging with that difference' (interview with author). In 2011, the piano was donated to the Medical Foundation for the Care of Victims of Torture (now Freedom From Torture) to be used in music therapy.

14 I visited the Biennale in connection with research for this project in 2013 and 2017. I also revisited the 2013 project when it was exhibited at the South London Gallery in Peckham in 2014, the only time that the Iraq Pavilion has been brought to Britain since 2011. As well as these visits, at which I took notes and photographs, this section draws on the catalogues published for each Pavilion, the curatorial and critical essays they contained, the website for the 2017 exhibition and other interviews, articles, critiques and polemics already in the public domain.

15 In 1989, Ahmad Chalabi had been convicted in absentia of bank fraud by a military court in Jordan.

Chapter Nine
Conclusions

Shortly after 11 am on 6 July 2016, Sir John Chilcot began to address an audience assembled in the Queen Elizabeth II Centre in Westminster, just around the corner from the Houses of Parliament and the main offices of government, before going on to summarise the findings of the public inquiry that he had chaired since being appointed by Prime Minister Gordon Brown in June 2009. As he recounted, Chilcot had been asked to investigate 'whether it was right and necessary to invade Iraq in March 2003; and, whether the UK could – and should – have been better prepared for what followed' (Iraq Inquiry 2016). Chilcot went on to outline findings that were far more critical of Tony Blair, his government and the civilian and military leadership of the invasion and occupation than many had anticipated, concluding that 'military action at that time was not a last resort', that intelligence judgements had been 'presented with a certainty that was not justified', that 'planning and preparations ... were wholly inadequate', and that '[t]he Government failed to achieve its stated objectives' (Iraq Inquiry 2016). Many family members of British service personnel killed in Iraq were in the audience, and, while Chilcot noted that the report had not been written for them, he acknowledged the importance of their presence. As he later described (BBC News 2017), family members had applauded him before and after he gave a short speech to them in private earlier that morning.

A public inquiry represents what I have called a technique of the event, and the launch of the report was itself an event, which brought Britain's Iraq war back into the public domain, intensifying feelings, orienting practices and eliciting

Geopolitics and the Event: Rethinking Britain's Iraq War Through Art, First Edition. Alan Ingram.
© 2019 Royal Geographical Society (with the Institute of British Geographers).
Published 2019 by John Wiley & Sons Ltd.

statements about the event with which it sought to reckon, thereby extending and adding further layers to it. The report reconstituted the event of the war as something different from what its architects had insisted it to be, in a manner that accorded more closely with the views of many observers. What the official transcript of the launch fails to record, however, is that, after thanking the staff of the conference centre, Chilcot had said:

> Before I begin my statement, perhaps we should all recall the continuing suffering of innocent people in Iraq, and those who have been killed and injured in terrorist attacks, including the latest horrific attack last Sunday, which killed more than two hundred and fifty people. (BBC News Channel 2016)

With this brief sentence, Chilcot acknowledged, but could not begin to encompass, the scale and nature of the losses that had been experienced by Iraqi people, and which they continued to endure. While the marginal consideration offered by the report to the effects of the war on the country and people of Iraq was a product of its narrow parameters, those parameters served to perpetuate a particular distribution of sensibility, in relation to which the 2003 Iraq war appears primarily as a problem of, and for, Britain, for the British government and military, and for British people more generally. Britain's Iraq war certainly needs to be scrutinised in these terms, but account must also be taken of how, in the course of sensing and making sense of the war in Britain, Iraqi people and Iraq itself often fail to appear, speak or be heard and of how, when they do appear, this tends to be in abbreviated, reductive or stereotypical ways.[1]

If they have remained in the margins of exercises like the Iraq Inquiry, the experiences as well as creative and critical work of people of Iraqi heritage concerning the 2003 war have been much more visible in the field of art, albeit in uneven and highly mediated ways. Though much of Iraq's cultural heritage was damaged, destroyed and stolen in the wake of the 2003 invasion, while Iraqi artists and intellectuals have been killed, injured and displaced, and while cultural institutions and cultural life have been dramatically curtailed and distorted in the country, artists, curators and writers of Iraqi heritage in Iraq and in the diaspora have continued to make work that engages with ongoing events and to make their work public in Britain and beyond, in ways that invite and demand the engagement of anyone trying to make sense of the event. The book is in significant part a response to their work and the demands that it makes.

This kind of engagement, I have argued, leads us towards an understanding of geopolitical events like the 2003 Iraq war as multiple, as well as singular, things. In this sense, a geopolitical event, conceptualised as a disruptive transformation of the world and of ways of sensing and making sense of it, is understood as being enacted, experienced and appropriated differently by different people, in different places, at different times, in ways that may challenge our sense of it being 'an' event at all. Britain's Iraq war is not equivalent to the war in Iraq itself or as experienced by people of Iraqi heritage in the diaspora, but these events are entangled with each other and might, via an appreciation of their multiplicity, be understood to constitute something like 'an' event.

An engagement with art enhances an appreciation of the manner in which geopolitical events are made up of, and form part of, other events, and of how the relations between these events can be constructed in a variety of ways. Also enhanced is a sense of how, while we can try to track, and assemble accounts of events as happening 'in' space and time, the temporalities and spatialities of an event are rarely straightforwardly linear or topographic, but rather broken and recursive, layered and twisted, such that events are liable to extend, recur, rhyme, echo and resonate with each other, and to endure in a spectral manner that troubles attempts to transcend, disavow or move beyond them.

Art is a way of enacting and experimenting with events, which, as I have argued, differs in important respects from procedures such as those followed by the Iraq Inquiry, by academic analysis or by protest. As IR theorist Roland Bleiker (2009) has argued, the value of an engagement with art is precisely that it does not offer something that is directly analytical or political, but indirectly so. I have argued that art is able to do this to the extent that it constitutes 'a specific sensorium that stands out as an exception from the normal regime of the sensible' (Rancière 2002, p.134, fn.1). As Rancière argues, art is a *dispositif* that enables a form of fiction. As he explains, fiction

> is not a term that designates the imaginary as opposed to the real; it involves the reframing of the 'real', or the reframing of a dissensus. Fiction is a way of changing existing modes of sensory presentations and forms of enunciation; of varying frames, scales and rhythms; and of building new relationships between reality and appearance, the individual and the collective. (2010, p.141)

The specificities of particular art forms and artworks matter and make a difference, but the understanding of art as a *dispositif* that enables artworks to do what they are widely held to do is something that has sometimes been under-appreciated in existing literature in political geography and geopolitics.

As I have argued, the *dispositif* in relation to which art can appear in a quasi-autonomous manner and be said to counter-actualise events is itself historically and geopolitically constituted in relation to nation-state and colonial formations of power and the related bifurcation between art and artefact. While themselves in motion, these relations and formations create positions of privilege and marginality, access and exclusion. But while the fields within and through which art plays out are not equitable, neither are they exclusive, nor hermetic. At the same time, influential players in art worlds and art markets have come at times to privilege and commodify art that seems to be representative of other things, including war, violence and trauma, or particular places, peoples, regions and cultures. For artists, this creates incentives to present oneself and one's art in terms of its representative qualities, that is, for example, as 'Iraqi' or 'war' art. While it might be said that this book has capitalised upon the effects of the resultant structures, I do not believe that the extensive body of work to have been created out of the 2003 Iraq war can be reduced to them, even where some artists or exhibitions may have been shaped, elevated and branded by commercial interests eager to capitalise on the timeliness of Iraq war art.

A related point is that I have encountered and researched art emerging in relation to Britain's Iraq war via a particular positionality and relation to British geopolitical culture and to colonial, patriarchal and anthropocenic geopolitics. In keeping with a genealogical spirit of inquiry, my goal has been to question and unsettle this culture and these forms of geopolitics, while recognising the difficulties of doing this from within, with conceptual tools and via a *dispositif* of art that are themselves implicated in the problem under study. A further issue with which the book has grappled is the inherently paradoxical activity of writing and thinking conceptually about something that is interesting precisely to the extent that it resists such appropriation. In so doing, I have tried to be faithful to the idea that art works as, variously and not exclusively, shock, invitation, prompt or compulsion to thought, but cannot be reduced to or encompassed within it. The existence of the book might therefore itself be taken as evidence of art's capacity to affect and to move, and while this has an inevitable subjective dimension, I hope that my explicit and implicit methodological and analytical choices and biases are apparent.

The book has advanced three main arguments. Taking inspiration from work on art in relation to the event, from philosophical, postcolonial and decolonial accounts of the event, and from the ways in which the event has been explored in political geography, international relations and related fields, I have argued, first, for developing the idea of the event as a fundamental focus for critical research on geopolitics. This argument has been driven to a large extent by the difficulty of thinking the 2003 Iraq war through existing ideas of the event, thus recognising its problematising character as an event, and informing my framing of the geopolitical event as a disruptive transformation of the world and of ways of sensing and making sense of it. Trying to approach and understand the 2003 war through art has also informed my understanding of it as an event that is made up of events, while forming part of other events, and as being a multiple as well as singular thing. Events can be said simply to happen, and to be experienced and undergone, while also being enacted or 'done', that is, engendered and altered, by practices and actions. Understanding events as multiple means recognising that these dimensions or qualities of an event, including the capacity to initiate, shape, avoid, survive or profit from them, are differentially distributed, in ways that shift over time and space. This recognises the phenomenological dimensions to events, while allowing the possibility of an event (or Event) that is not reducible to individual or collective experience, and which is more than is actualised or experienced empirically. Furthermore, as Paul Patton's

(1997) work on the event and Andrew Barry's (2013) conceptualisation of the political situation would suggest, the nature, parameters and even existence of a geopolitical event are invariably part of what is at stake in an event, over which parties to the event may struggle and disagree. As some literature in political geography (de Goede 2014) and critical security studies (Walters 2014) has started to explore, and as the book has explored, establishing accounts of events is a pre-eminently political activity. As I have argued, approaching events genealogically raises questions of ontology and epistemology, as well as ethics, politics and aesthetics, that are central to geopolitics as a field of inquiry.

I have argued, second, for a revised and expanded engagement with geopolitical events via art. This, I hope to have shown, is particularly valuable in terms of Britain's Iraq war, but is also relevant to other events that have generated artworks. I have suggested that we can think about art and geopolitics through each other by conceptualising artworks as evental assemblages, that is, as open, heterogeneous things involving material, energetic, embodied and semiotic elements, sensation and sense-making, which emerge from events, themselves constitute events, and which may engender and participate in ongoing events. This has methodological implications, in that it is necessary to inquire into the ways in which artworks are created; their formal properties; their relations to their institutional and discursive contexts; the ways they may come into public domains and assemble temporary publics (or fail to do so); and the ways in which they may be said to participate in ongoing events.

Approaching art and geopolitics in evental terms, I suggest, helps to avoid two methodological pitfalls: reducing works to a biographical analysis (something which I believe is offset by examining many works across an event) or seeking to establish the reception or 'meaning' of a work by asking people what they think about it (which I believe is offset by engaging non-directively with the communities of sense through which works emerge and tracking public responses to them across an extended period of time). What I hope to have added to existing work on art in relation to geopolitics and international relations is a more genealogical sense of art itself as a *dispositif* and a set of orientations for exploring events as they are counter-actualised across many artworks across and through time and space, while recognising the forms of time and space that they themselves incarnate. Rather than arguing that a specific art form is particularly suited to counter-actualisation, or trying to identify artworks that are particularly emblematic of this, I have turned the problem around and proceeded from the position that, if something has been accepted as an artwork relating to the war, that is, if it is already positioned in relation to the *dispositif* of art as well as the event, then it in some sense is already a counter-actualisation of the event. To engage with art is then also to engage with the virtual dimensions of geopolitics. This allows a backgrounding of problems of judgement, such as whether a given art work is 'good' or 'successful' in aesthetic, ethical or political terms. The prior issue concerns the ways in which a given entity comes to be 'art' at all.

Third, I have argued for rethinking a specific event, the 2003 Iraq war, or still more specifically, Britain's Iraq war, through art. Little academic work to date has examined this war through the art that has emerged from it. To the extent that academic and curatorial projects have addressed the war, they have with few exceptions (and often with good reason) tended to do so in terms of a particular community of sense, rather than taking the existence of distinct communities of sense in relation to the event as something to be recognised or investigated. Existing academic work gives us ways to think critically about the 2003 Iraq war in relation to, *inter alia*, a new imperialism (Harvey 2003), military neoliberalism (Boal *et al.* 2003) and colonial imaginaries (Gregory 2004a, 2004b), while a closer examination of journalistic and personal accounts (e.g. Stewart 2006; Steele 2009; Fairweather 2011; Ledwidge 2011) would no doubt yield further insights as to the specificities of British geopolitical cultures (Toal 2008) with regard to the war.[2] What a sustained focus on diverse artworks emerging from the war reveals is a field in which specific geopolitical ideologies, imaginaries, technologies and practices have been appropriated, altered and experimented upon in a critical and creative manner as part of – but also, fleetingly and provisionally, autonomously from – the ongoing becoming of the world. If we want to understand how geopolitics is assembled and how it works in the interstices of affect and embodiment, technology and materiality, cognition and practice, sensation and sense-making, then an engagement with art practice, which has long been concerned with precisely such matters, is imperative. Art offers something to genealogical inquiry into the 2003 Iraq war in particular in terms of how it offers circuitous, indirect, materially-mediated pathways – and, sometimes, obstacles and diversions – to a cognitive appreciation of an event that other techniques have left unresolved, and which continues to affect political life in Britain as well as Iraq and the region.

Britain's Iraq wars continue to be enacted and actualised in multiple ways in relation to other geopolitical events, and while academics and institutions from Britain continue to be involved in reparative and reconstructive efforts in relation to Iraq's cultural and archaeological heritage, new sites for remembrance have been established in Britain. In 2010, a memorial originally constructed at a British base in Southern Iraq to commemorate the deaths of British service personnel and contractors working with them was reassembled at the National Memorial Arboretum in Staffordshire, itself a recently established site for military commemoration, and a ceremony of rededication took place there. The 'Basra Wall' now stands amid memorials to numerous other military interventions, as a permanent reminder of British military and associated losses. In March 2017, a new Iraq and Afghanistan memorial, located in Victoria Embankment Gardens next to the Ministry of Defence in London, was dedicated by Queen Elizabeth II. Reflecting the difficulty of establishing the parameters of the events that it commemorates, a government statement announced that the memorial 'recognises the contributions of the UK Armed Forces and all UK citizens who deployed in the Gulf region, Iraq and Afghanistan from 1990–2015, and those who supported them back home' (GOV.UK 2017). As the statement concluded, '[t]he combined events that the memorial covers represent the longest post-war continuous overseas deployment of UK forces (excepting garrison duties), and the most intensive extended period of operations undertaken since the Second World War' (GOV.UK 2017). The memorial and its accompanying statements thus offer some insight into the multiplicity of geopolitical events as well as the ongoing, complicated geopolitics of Britain in the world, but in relation to only one, clearly delimited, community of sense. While Chilcot had found a moment to acknowledge Iraqi losses, neither the memorial nor the brochure published to accompany its dedication (Ministry of Defence 2017) make any such mention.

It is therefore worth returning to the *lamassu* that appeared for a time on the Fourth Plinth in London's Trafalgar Square and which memorialised the event in a different way. Positioned in the corner of the Square, the *lamassu* stood with its back to the National Gallery, facing more or less southwards and gazing intently over the heads of visitors and beyond Nelson's Column, down Whitehall, towards Number 10, Parliament, the Foreign and Commonwealth Office and the Ministry of Defence, and towards the many commemorative objects that surround them, including the Iraq and Afghanistan memorial (Figure 9.1).

Figure 9.1   *The Invisible Enemy Should Not Exist*, Michael Rakowitz (2018). A Mayor of London Fourth Plinth Commission (alternate view). Photo GLA/Caroline Teo, courtesy of the artist.

If you are able to visit the British Museum, one of the guides who conducts tours of its Mesopotamian Galleries will tell you of how, in constructing the *lamassu*, Assyrian sculptors sought to materialise the most powerful figure imaginable, combining the head of a king with the wings of an eagle, and the bodies, legs and feet of bulls and lions. Although the *lamassu* embodied and projected the power of imperial civilisations, which periodically expanded to dominate surrounding peoples and regions, they are also understood to have been protective deities.

As a statement issued to mark the unveiling of the sculpture read, '[r]ebuilding the Lamassu in Trafalgar Square means it can continue to guard the people who live, visit and work in London' (Greater London Authority 2018). This statement appears to appropriate the *lamassu* as being for London and at home there; while its double can no longer perform its role in Iraq, we might surmise, the reproduction can here. In some sense, we might infer, London becomes a secure home for the object lost in Iraq, perhaps echoing a discourse within which Iraq's cultural heritage is not safe within the country and is better located overseas. We might even take it, approvingly or not, to be an appropriate symbol for an imperial nation that has much to guard. As the caption on the Plinth itself reads, however, '[r]ebuilding the Lamassu means it can symbolically continue as guardian of *a* city's *past, present and future*' (emphasis added). This difference in framing opens the sculpture up to events in subtly different ways, emphasising the past as well as the present and future. It is also less clear from the reference to 'a' city that this reconstructed *lamassu* is 'for' London, or at home there. It is not just that any city might do, however, nor that the *lamassu* might haunt as well as guard, but that such an object should be at home in any city, whose past, present and future ought, like any other, to be protected. This reading hints at how we can understand the work as a counter-actualisation, as an event that emerges from, and refers to, the ongoing, evental, actualisation of geopolitics, while making something else of things and pointing, ambiguously and against all actuality, towards another kind of world.

Notes

1   This problem was related forcefully to me by several interviewees with connections to Iraq.
2   An extensive body of work by Rachel Woodward, Neil Jenkings and co-authors has explored the photographs and memoirs composed by British soldiers: e.g. Woodward and Jenkings (2012); Woodward and Jenkings (2016).

References

Adey, P., Anderson, B. and Lobo Guerrero, L. (2011). An ash cloud, airspace and environmental threat. *Transactions of the Institute of British Geographers* 36 (3): 338–343.

Adey, P., Whitehead, M. and Williams, A.J. (2013). *From Above: War, Violence and Verticality*. London: Hurst.

Adorno, T. (1974). Commitment. *New Left Review I* (87–88): 75–89.

Adorno, T. (2013). *Aesthetic Theory*. London: Bloomsbury.

Adorno T., Benjamin, W., Bloch, E., Brecht, B. Lukács, G. and Jameson, F. (2007). *Aesthetics and Politics*. London: Verso.

Agamben, G. (1998). *Homo Sacer: Sovereign Power and Bare Life*. Stanford, CA: Stanford University Press.

Agamben, G. (2000). *Remnants of Auschwitz: The Witness and the Archive*. New York: Zone Books.

Agamben, G. (2005). *State of Exception*. London: University of Chicago Press.

Agnew, J. (1998). *Geopolitics: Re-Visioning World Politics*. London: Routledge.

Ahmed, O. (2007). *Displaced*. London: Imperial War Museum.

Al-Ali, N. (2007). *Iraqi Women: Untold Stories from 1948 to the Present*. London: Zed Books.

Al-Ali, N. (2011). A feminist perspective on the Iraq war. *Works and Days* 57/58 (29): 1–15.

Al-Ali, N., and Al-Najjar, D. (2012). *We Are Iraqis: Aesthetics and Politics in a Time of War*. Syracuse NY: Syracuse University Press.

Al-Ali, N., and Pratt, N. (2009). *What Kind of Liberation? Women and the Occupation of Iraq*. Berkeley: University of California Press.

Al-Gailani Werr, L. (2013). Archaeology and politics in Iraq. In: *Critical Approaches to Near Eastern Art* (eds. B.A. Brown and M.H. Feldman), 15–30. Berlin: de Gruyter.

Al-Gailani Werr, L. (2014). World War I and archaeology in Iraq. http://asorblog. org/2014/07/07/world-war-i-and-archaeology-in-iraq/ (accessed 20 December 2017).

Allawi, A. (2007). *The Occupation of Iraq: Winning the War, Losing the Peace*. New Haven, CT: Yale University Press.

Geopolitics and the Event: Rethinking Britain's Iraq War Through Art, First Edition. Alan Ingram.
© 2019 Royal Geographical Society (with the Institute of British Geographers).
Published 2019 by John Wiley & Sons Ltd.

Allen, J. (2011). Topological twists: power's shifting geographies. *Dialogues in Human Geography* 1 (3): 283–298.

Alphen, E. van. (2005). *Art in Mind: How Contemporary Images Shape Thought*. Chicago, IL: University of Chicago Press.

Alÿs, F. (2017). Francis Alÿs: Embedded with the Peshmerga. In: *Archaic: the Pavilion of Iraq: 57th International Art Exhibition La Biennale di Venezia* (ed. S. Cernuschi), 48–51. Milan: Mousse.

Amoore, L. (2006). Biometric borders: governing mobilities in the War on Terror. *Political Geography* 25 (3): 336–351.

Amoore, L. (2013). *The Politics of Possibility: Risk and Security Beyond Probability*. Durham NC: Duke University Press.

Amoore, L., and Hall, A. (2010). Border theatre: on the arts of security and resistance. *Cultural Geographies* 17 (3): 299–319.

Anderson, Benedict. (1983). *Imagined Communities: Reflections on the Origin and Spread of Nationalism*. London: Verso.

Anderson, Ben. (2010). Preemption, precaution, preparedness: anticipatory action and future geographies. *Progress in Human Geography* 34 (6): 777–798.

Anderson, Ben. (2014). *Encountering Affect: Capacities, Apparatuses, Conditions*. Farnham: Ashgate.

Anderson, Ben, and Gordon, R. (2016). Government and (non)event: the promise of control. *Social & Cultural Geography* 18 (2): 158–177.

Anderson, Ben, and Harrison, P., eds. (2010). *Taking Place: Nonrepresentational Theory and Geography*. Farnham: Ashgate.

Antoon, S. (2003). Dead poets society. *The Nation*. https://www.thenation.com/article/dead-poets-society/ (accessed 5 November 2018).

Aradau, C., and Munster, R. (2007). Governing terrorism through risk: taking precautions, (un)knowing the future. *European Journal of International Relations* 13 (1): 89–115.

Arango, T. (2016). After 25 years of US role in Iraq, scars are too stubborn to fade. *The New York Times* (16 February). https://www.nytimes.com/2016/02/17/world/middleeast/25th-anniversary-of-us-involvement-passes-quietly-for-iraqis-unsure-of-future.html (accessed 13 July 2017).

Armstrong, I. (2000). *The Radical Aesthetic*. Oxford: Blackwell.

Art Exchange. (2009). Queen and Country: a project by Steve McQueen. http://www.artexchange.org.uk/exhibition/queen-and-country-a-project-by-steve-mcqueen (accessed 17 December 2017).

ArtRabbit (2010). Rashad Selim – Soundbase. 19 August 2010–26 September 2010. https://www.artrabbit.com/events/rashad-salim-soundbase (accessed 15 December 2017).

ArtRole (2009). Post War Festival. http://artrole.org/projects/post-war-festival/ (accessed 18 January 2019).

Asfahani, R. (2016). Interview with Hanaa Malallah: Iraq's pioneering female artist. Web page. https://theculturetrip.com/middle-east/iraq/articles/interview-with-hanaa-malallah-iraq-s-pioneering-female-artist/ (accessed 18 April 2018).

Aspden, P. (2007). McQueen and Country. *Financial Times* (28 February), p.12.

Attwood, P., and Powell, F. (2009). *Medals of Dishonour*. London: British Museum.

Azoulay, A. (2008). *The Civil Contract of Photography*. New York: Zone Books.

Badiou, A. (2012). *The Rebirth of History: Times of Riots and Uprisings*. London: Verso.

Badiou, A. (2013). *Logic of Worlds: Being and Event II*. London: Bloomsbury.

Badr, A., Rashid, A., Harram, F., MacKinnon, K., Gailani Werr, L., Luk, L., Cauteren, P. van, Kealy, S., Bekes and S., Chalabi, T. (2015). Invisible Beauty: the Iraq Pavilion at the 26th international art exhibition, la biennale di Venezia. Milan: Mousse; Baghdad: Ruya Foundation.

Bahrani, Z. (2001). *Women of Babylon: Gender and Representation in Antiquity*. London: Routledge.

Bahrani, Z. (2003). Looting and conquest. *The Nation* (14 May). http://www.thenation.com/article/looting-and-conquest (accessed 3 August 2012).

Bahrani, Z. (2008). *Rituals of War: The Body and Violence in Mesopotamia*. New York: Zone Books.

Bahrani, Z. (2009). Modernism and Iraq. In: *Modernism and Iraq* (eds. Z. Bahrani and N. Shabout), 11–22. New York: Miriam and Ira D. Wallach Art Gallery, Columbia University.

Bahrani, Z. (2013). Regarding art and art history. *The Art Bulletin* 95 (4): 516–517.

Bahrani, Z. (2014). *The Infinite Image: Art, Time and the Aesthetic Dimension in Antiquity*. London: Reaktion.

Bahrani, Z. (2017). The archaic aesthetic. In: *Archaic: the Pavilion of Iraq: 57th International Art Exhibition La Biennale di Venezia* (ed. S. Cernuschi), 12–16. Milan: Mousse.

Bahrani Z., and Shabout N. (2009). *Modernism and Iraq*. New York: Miriam and Ira D. Wallach Art Gallery, Columbia University.

Bailey, J., Iron, R. and Strachan, H., eds. (2014). *British Generals in Blair's Wars*. London: Routledge.

Barjeel Art Foundation. (2017). Layla Al-Attar. http://www.barjeelartfoundation.org/artist/iraq/layla-al-attar/ (accessed 21 December 2017).

Barker, K. (2012). Influenza preparedness and the bureaucratic reflex: anticipating and generating the 2009 H1N1 event. *Health and Place* 18 (4): 701–709.

Barr, J. (2016). Sykes-Picot is not to blame for Middle East's problems. *Al Jazeera* (16 May). http://www.aljazeera.com/indepth/opinion/2016/05/sykes-picot-blame-middle-east-problems-160516083606123.html (accessed 15 December 2017).

Barragán, P. (2012). Interview with Martha Rosler. *Artpulse* 12 (Summer): 48–51.

Barringer, T., and Flynn, T. (1998). *Colonialism and the Object: Empire, Material Culture and the Museum*. London: Routledge.

Barry, A. (2013). *Material Politics: Disputes Along the Pipeline*. Oxford: Wiley-Blackwell.

Bassett, K. (2016). Event, politics and space: Rancière or Badiou? *Space and Polity* 20 (3): 280–293.

Batatu, H. (1978). *The Old Social Classes and Revolutionary Movements of Iraq: A Study of Iraq's Old Landed and Commercial Classes and of its Communists, Ba'thists and Free Officers*. Princeton NJ: Princeton University Press.

beaubodor. (2015). Tweet. (24 November). https://twitter.com/beaubodor/status/669174416723615745 (accessed 17 December 2017).

BBC. (2008). On this day: 17 June 1980: Government announces missile sites. http://news.bbc.co.uk/onthisday/hi/dates/stories/june/17/newsid_2514000/2514879.stm (accessed 17 December 2017).

BBC News. (2003a). Interest grows in Iraqi art. (31 March). http://news.bbc.co.uk/1/hi/entertainment/2902033.stm (accessed 15 December 2017).

BBC News. (2003b). Transcript of Blair's Iraq Interview. (6 February). http://news.bbc.co.uk/1/hi/programmes/newsnight/2732979.stm (accessed 6 Devember 2017).

BBC News. (2004). Army photos: claims and rebuttals. (14 May). http://news.bbc.co.uk/1/hi/uk/3680327.stm (accessed 17 December 2017).

BBC News. (2005a). British troops arrested in Basra. (19 September). http://news.bbc.co.uk/1/hi/world/middle_east/4260894.stm (accessed 6 December 2017).

BBC News. (2005b). London bomber: Text in full. (1 September). http://news.bbc.co.uk/1/hi/uk/4206800.stm (accessed 13 December 2017).

BBC News. (2005c). Two soldiers guilty of Iraq abuse. (23 February). http://news.bbc.co.uk/1/hi/uk/4290435.stm (accessed 17 December 2017).

BBC News. (2006a). Huge gaps between Iraq death estimates. (20 October). http://news.bbc.co.uk/1/hi/world/middle_east/6045112.stm (accessed 14 December 2017).

BBC News. (2006b). Video of 7 July bomber released. (6 July). http://news.bbc.co.uk/1/hi/5154714.stm (accessed 13 December 2017).

BBC News. (2013). Row over Tony Blair 'selfie' artwork. (18 October). http://www.bbc.co.uk/news/entertainment-arts-24565194 (17 December 2017).

BBC News. (2017). In full: Laura Kuenssberg interview with Sir John Chilcot. Video. http://www.bbc.co.uk/news/av/uk-politics-40516908/in-full-laura-kuenssberg-interview-with-sir-john-chilcot (accessed 11 April 2018).

BBC News Channel. (2016). Iraq Inquiry Report. https://www.youtube.com/watch?v=NCUOHOMNq_A (accessed 11 April 2018).

BBC World Service. (2009). The Strand. (25 September). http://www.bbc.co.uk/programmes/p004bkgm (accessed 15 December 2017).

Bell, G. (1927). *The Letters of Gertrude Bell. Selected and Edited by Lady Bell, DBE Volume 2.* New York: Boni and Liveright.

Bennett, Jane. (2010). *Vibrant Matter: A Political Theory of Things.* Durham NC: Duke University Press.

Bennet, Jill. (2005). *Empathic Vision: Affect, Trauma and Contemporary Art.* Stanford, CA: Stanford University Press.

Bennett, Jill. (2012). *Practical Aesthetics: Events, Affects and Art after 9/11.* London: I.B. Tauris.

Bennett, T. (1988). The exhibitionary complex. *New Formations* 4 (1): 73–102.

Bennett, T. (2015). Thinking (with) museums: from exhibitionary complex to governmental assemblage. In: *The International Handbooks of Museum Studies: Museum Theory* (eds. A. Witcomb and K. Message), 3–20. Oxford: Wiley-Blackwell.

Berlant, L. (2011). *Cruel Optimism.* Durham NC: Duke University Press.

Berman, M. (1983). *All That Is Solid Melts Into Air: The Experience of Modernity.* London: Verso.

Bernhardsson, M.T. (2005). *Reclaiming a Plundered Past: Archaeology and Nation Building in Modern Iraq.* Austin, TX: University of Texas Press.

Bevan, S. (2015). *Art from Contemporary Conflict.* London: Imperial War Museums.

Bilal, W., and Lydersen, K. (2008). *Shoot an Iraq: Art, Life and Resistance Under the Gun.* San Francisco, CA: City Lights.

Billig, M. (1995). *Banal Nationalism.* London: Sage.

Biro, M. (2009). *The Dada Cyborg: Visions of the New Human in Weimar Berlin.* Minneapolis: University of Minnesota Press.

Bishop, C. (2010). *Installation Art: A Critical History.* London: Tate.

Blair, T. (2007). Blair's speech on the arts in full. *The Guardian* (6 March). https://www.theguardian.com/politics/2007/mar/06/politicsandthearts.uk1 (accessed 17 December 2017).

Blanchot, M. (1995). *The Writing of the Disaster*. London: University of Nebraska Press.

Blazwick, I. (2008). Taking responsibility: Martha Rosler interviewed by Iwona Blazwick. *Art Monthly* 314 (March), 1–7.

Bleiker, R. (1998). Retracing and redrawing the boundaries of events: postmodern interferences with international theory. *Alternatives* 23 (4): 471–497.

Bleiker, R. (2009). *Aesthetics and World Politics*. Houndmills: Palgrave Macmillan.

Boal, I., Clark, T.J., Matthews, J. and Watts, M. (2005). *Afflicted Powers: Capital and Spectacle in a New Age of War*. London: Verso.

Bogdanos, M. (2005). The casualties of war: the truth about the Iraq Museum. *American Journal of Archaeology* 109: 477–526.

Bohrer, F. (1998). Inventing Assyria: exoticism and reception in nineteenth-century England and France. *The Art Bulletin* 80 (2): 336–356.

Bois, Y.-A., Brett, G., Iversen, M. and Stallabrass, J. (2008). An interview with Mark Wallinger. *October* 123: 185–204.

Bonneuil, C., and Fressoz, J.-B. (2017). *The Shock of the Anthropocene*. London: Verso.

Borradori, G., Habermas, J. and Derrida J. (2003). *Philosophy in a Time of Terror: Dialogues with Jurgen Habermas and Jacques Derrida*. Chicago, IL: Chicago University Press.

Braidotti, R. (2013). *The Posthuman*. Cambridge: Polity.

Brickell, K. Geopolitics of home. *Geography Compass* 6 (10): 575–588.

British Museum. (2008). *Iraq's Past Speaks to the Present*. Exhibition catalogue. London: British Museum.

British Museum. (2017). Boushra Almutawakel to Michael Rakowitz: Recent acquisitions of works by Arab artists. http://www.britishmuseum.org/whats_on/exhibitions/arab_art. aspx (accessed 17 December 2017).

Brookes, P. (2015). Tweet. (8 April). https://twitter.com/brookestimes/status/ 585774276764180480 (accessed 17 December 2017).

Brown, H., Maunsell, Colonel, Bell, G., Moncrieff C.S. and Willcocks, W. (1910). Mesopotamia: past, present and future: discussion. *The Geographical Journal* 35 (1): 15–18.

Bryant, L., Srnicek, N. and Harman, G., eds. (2011). *The Speculative Turn: Continental Materialism and Realism*. Melbourne: re.press.

Buchanan, I., and Lambert, G., eds. (2005). *Deleuze and Space*. Edinburgh: Edinburgh University Press.

Burch, S. (2017). A virtual oasis: Trafalgar Square's arch of Palmyra. *International Journal of Architectural Research* 11 (3): 58–77.

Burke, G. (2007). *The National Theatre of Scotland's Black Watch*. London: Faber and Faber.

Burkeman, O. (2005). First mission: to turn the tide in Dorset. *The Guardian* (6 April). https:// www.theguardian.com/politics/2005/apr/06/uk.election20052 (accessed 17 December 2017).

Butler, J. (2004). *Precarious Life: The Powers of Mourning and Violence*. London: Verso.

Butler, J. (2010). *Frames of War: When is Life Grievable?* London: Verso.

Campbell, D. (2001). Time is broken: the return of the past and the response to September 11. *Theory and Event* 5 (4). DOI: 10.1353/tae.2001.0032.

Caplan, C. (2013). Sensing distance: the time and space of contemporary war. *Social Text Online* (17 June). https://socialtextjournal.org/periscope_article/sensing-distance-the-time-and-space-of-contemporary-war/ (accessed 4 January 2018).

Carden-Coyne, A. (2009). *Reconstructing the Body: Classicism, Modernism, and the First World War*. Oxford: Oxford University Press.

Casati, R., and Varzi, A., eds. (1996). *Events*. Aldershot: Dartmouth.

Cavarero, A. (2011). *Horrorism: Naming Contemporary Violence*. New York: Columbia University Press.

Cernuschi, S., ed. (2017). *Archaic: the Pavilion of Iraq: 57th International Art Exhibition La Biennale di Venezia*. Milan: Mousse.

Chandrasekaran, R. (2007). *Imperial Life in the Emerald City: Inside Baghdad's Green Zone*. London: Bloomsbury.

Chulov, M. (2014). ISIS: the inside story. *The Guardian* (11 December). https://www.theguardian.com/world/2014/dec/11/-sp-isis-the-inside-story accessed 15 December 2017.

Clark, N. (2014). Geo-politics and the disaster of the Anthropocene. *The Sociological Review* 62: 19–37.

Clunas, C. (1994). Oriental antiquities/Far Eastern art. *Positions* 2 (2): 318–355.

Cockburn, P. (2006). *The Occupation: War and Resistance in Iraq*. London: Verso.

Cole, A. (2015). Those obscure objects of desire. *Artforum* 53 (10): 319–323.

Colebrook, C. (2010). Actuality. In: *The Deleuze Dictionary* (ed. A. Parr), 9–11. Edinburgh: Edinburgh University Press.

Colwell, C. (1997). Deleuze and Foucault: series, event, genealogy. *Theory & Event* 1 (2): DOI 10.1353/tae.1997.0004.

Copeland, H. (2017). Red, black and blue: the National Museum of African American History and Culture and the National Museum of the American Indian. *Artforum*. 56 (1): 253–261.

Costs of War. (2018). About. http://watson.brown.edu/costsofwar/about (accessed 11 April 2018).

Crawford, N. (2013). Civilian death and injury in the Iraq War, 2003–2013. watson.brown.edu/costsofwar/files/cow/imce/papers/2013/Civilian%20Death%20and%20Injury%20in%20the%20Iraq%20War%2C%202003-2013.pdf (accessed 11 April 2018).

Curtis, J. (2004). Report on Meeting at Babylon 11–13 December 2004. https://www.britishmuseum.org/PDF/BabylonReport04.pdf (accessed 15 December 2017).

Curtis, J. (2009). Relations between archeologists and the military in the case of Iraq. *Papers from the Institute of Archaeology* 19: 2–8.

Dalby, S. (2007). Anthropocene geopolitics: globalisation, empire, environment and critique. *Geography Compass* 1 (1): 103–118.

Dalby, S. (2016). Framing the Anthropocene: the good, the bad and the ugly. *The Anthropocene Review* 3 (1): 33–51.

Dalby, S. (2017). Firepower: geopolitical cultures in the anthropocene. *Geopolitics* https://doi.org/10.1080/14650045.2017.1344835.

Danchev, A. (2011). *On Art and War and Terror*. Edinburgh: Edinburgh University Press.

Danto, A. (1964). The artworld. *The Journal of Philosophy* 61 (19): 571–584.

Davis, A.J. (2013). Star Wars: return of the sixties, or, Martha Rosler versus the Empire striking back. *Third Text* 27 (4): 565–578.

Davis, E. (2005). *Memories of State: Politics, History, and Collective Identity in Modern Iraq*. Berkeley: University of California Press.

Davis, H., and Todd, Z. (2017). On the importance of a date, or, decolonizing the Anthropocene. *ACME: An International Journal for Critical Geographies* 16 (4): 761–780.

Davis, H., and Turpin, E., eds. (2015). *Art in the Anthropocene: Encounters Among Aesthetics, Politics, Environments and Epistemologies*. London: Open Humanities Press.

Dear, M., Ketchum, J., Luria, S. and Richardson, D., eds. (2011). *Geohumanities: Art, History, Text at the Edge of Place*. London: Routledge.

Debrix, F. (2008). *Tabloid Terror: War, Culture and Geopolitics*. London: Routledge.

DeLanda, M. (1997). *A Thousand Years of Nonlinear History*. New York: Zone Books.

DeLanda, M. (2013). *Intensive Science and Virtual Philosophy*. London: Bloomsbury.

Deleuze, G. (2004). *Difference and Repetition*. London: Continuum.

Deleuze, G. (2005). *Francis Bacon: The Logic of Sensation*. London: Continuum.

Deleuze, G. (2015). *The Logic of Sense*. London: Bloomsbury.

Deleuze, G., and Guattari, F. (1994). *What is Philosophy?* London: Verso.

Deleuze, G., and Guattari, F. (2004). *A Thousand Plateaus: Capitalism and Schizophrenia*. London: Continuum.

Deller, J. (2010). *It Is What It Is*. New York: Creative Time.

Deller, J. (2012). *Joy in People*. London: Southbank Centre.

Demos, T.J. (2013). *Return to the Postcolony: Specters of Colonialism in Contemporary Art*. Berlin: Sternberg Press.

Demotix. (2011). Stop the War Coalition projection close to Chilcot Enquiry. https://www.youtube.com/watch?v=e-ub1qwnfbc (accessed 18 April 2018).

Derrida, J. (2007). A certain impossible possibility of saying the event. *Critical Inquiry* 33: 441–461.

De Vasconcellos, I. (2016). *Fourth Plinth: How London Created the World's Smallest Sculpture Park*. London: Art/Books.

Devetak, R. (2009). After the event: Don DeLillo's *White Noise* and September 11 narratives. *Review of International Studies* 35 (4): 795–815.

Dewachi, O. (2017). *Ungovernable Life: Mandatory Medicine and Statecraft in Iraq*. Stanford, CA: Stanford University Press.

Dewey, J. (2005). *Art as Experience*. New York: Penguin.

Dewsbury, J.D. (2007). Unthinking subjects: Alain Badiou and the event of thought in thinking politics. *Transactions of the Institute of British Geographers* 32 (4): 443–459.

Dewsbury, J.D., and Thrift, N. (2005). Genesis eternal: after Paul Klee. In: *Deleuze and Space* (eds. I. Buchanan and G. Lambert), 89–108. Edinburgh: Edinburgh University Press.

Dickie, G. (1974). *Art and the Aesthetic: An Institutional Analysis*. (Ithaca, NY: Cornell University Press).

Dikeç, M. (2005). Space, politics and the political. *Environment and Planning D: Society and Space* 23: 171–188.

Dikeç, M. (2013). Beginners and equals: political subjectivity in Arendt and Rancière. *Transactions of the Institute of British Geographers* 38 (1): 78–90.

Dittmer, J. (2013). Geopolitical assemblages and complexity. *Progress in Human Geography* 38 (3): 385–401.

Dixon, D.P. (2009). Creating the semi-living: on politics, aesthetics and the more-than-human. *Transactions of the Institute of British Geographers* 34 (4): 411–425.

Dixon, D.P. (2015). *Feminist Geopolitics: Material States*. London: Routledge.

Dixon, D.P., and Marston, S.A., eds. (2013). *Feminist Geopolitics: At the Sharp End*. London: Routledge.

Dixon, D., Hawkins, H. and Straughan, E. (2012). Of human birds and living rocks remaking aesthetics for post-human worlds. *Dialogues in Human Geography* 2 (3): 249–270.

Dodge, D. (2003). *Inventing Iraq: The Failure of Nation Building and a History Denied*. New York: Columbia University Press.

Dodge, D. (2012). *Iraq: From War to a New Authoritarianism*. Abingdon: Routledge.

Doel, M. (2017). *Geographies of Violence: Killing Space, Killing Time*. London: Sage.

Dolberg, E. (2010). *Open Shutters Iraq*. London: Trolley Books.

Dowler, L., and Sharp, J. (2001). A feminist geopolitics? *Space and Polity* 5 (3): 165–176.

Eagleton, T. (1990). *The Ideology of the Aesthetic*. Oxford: Blackwell.

Edkins, J. (2003). *Trauma and the Memory of Politics*. Cambridge: Cambridge University Press.

Edkins, J. (2015). *Face Politics*. London: Routledge.

Eisteddfod. (2018). About us. https://eisteddfod.wales/about-us (accessed 4 April 2018).

Elbaor, C. (2017). Michael Rakowitz wins 2018 Fourth Plinth Commission. *Artnet News* (21 March). https://news.artnet.com/art-world/fourth-plinth-michael-rakowitz-heather-phillipson-897860 (accessed 1 December 2017).

Elden, S. (2009). *Terror and Territory: The Spatial Extent of Sovereignty*. London: University of Minnesota Press.

Esson, J., Noxolo, P., Baxter, R., Daley, P. and Byron, M. (2017). The 2017 RGS-IBG Chair's theme: decolonising geographical knowledges, or reproducing coloniality? *Area* 49 (3): 384–388.

Evening Standard. (2013). Tony Blair's crazed selfie included in Imperial War Museum's new Art and War exhibition. *Evening Standard* (9 October). https://www.standard.co.uk/news/london/tony-blairs-crazed-selfie-included-in-imperial-war-museums-new-art-and-war-exhibition-8869420.html (accessed 17 December 2017).

Fairweather, J. (2011). *A War of Choice: The British in Iraq 2003–9*. London: Jonathan Cape.

Fanon, F. (2001). *The Wretched of the Earth*. Harmondsworth: Penguin.

Faraj, M., ed. (2001). *Strokes of Genius: Contemporary Iraqi Art*. London: Saqi Books.

Faruqi, S., ed. (2011). *Art in Iraq Today*. Milan: Skira.

Fawn, R., and Hinnebusch, R., eds. (2006). *The Iraq War: Causes and Consequences*. London: Lynne Reinner.

Fernand, D. (2013). Art and minds. *The Sunday Times Magazine* (3 May), pp.42–47.

Filipovic, E., van Hal, M. and Øvstebø, S., eds. (2010). *The Biennial Reader*. Bergen: Bergen Kunsthall.

Fiske, C. (2013). Frustrating desires: Q + A with Martha Rosler. *Art in America* (14 January). http://www.artinamericamagazine.com/news-features/interviews/martha-rosler-moma/ (accessed 17 December 2017).

Flint, C. (2016). *Geopolitical Constructs: The Mulberry Harbours, World War Two, and the Making of a Militarized Transatlantic*. London: Rowman and Littlefield.

Foucault, M. (1980). The confession of the flesh. In *Power/Knowledge: Selected Interviews and Other Writings 1972–1977* (ed. C. Gordon), 194–228. Hemel Hempstead: Harvester Wheatsheaf.

Foucault, M. (1998). *The History of Sexuality Volume 1: The Will to Knowledge*. London: Penguin.

Foucault, M. (2000). Nietzsche, genealogy, history. In: *Michel Foucault: Aesthetics: Essential Works of Foucault 1954–1984. Volume 2*. (ed. J. Faubion), 223–238. London: Penguin.

Foucault, M. (2002). Questions of method. In: *Michel Foucault: Power: Essential Works of Foucault 1954–1984. Volume 3*. (ed. J. Faubion), 223–238. London: Penguin.

Foucault, M. (2007). *Security, Territory, Population. Lectures at the Collège de France 1977–1978*. Basingstoke: Palgrave Macmillan.

Gardner, A., and Green, C. (2013). Biennials of the South on the edges of the global. *Third Text* 27 (4): 442–455.

Gaztambide-Fernández, R. (2014). Decolonial options and artistic/aestheSic entanglements: an interview with Walter Mignolo. *Decolonization: Indigeneity, Education and Society* 3 (1): 196–212.

Getty Images. (2003). President Bush visits Blair at 10 Downing Street. (20 November). http://www.gettyimages.co.uk/detail/news-photo/united-states-president-george-w-bush-and-britains-prime-news-photo/525619086?esource=SEO_GIS_CDN_Redirect.#united-states-president-george-w-bush-and-britains-prime-minister-picture-id525619086 (accessed 17 December 2017).

Getty Images (2005). General Election 2005 – Campaign Launch – Labour Party. Photocollection.http://www.gettyimages.co.uk/detail/news-photo/britains-prime-minister-tony-blair-looks-at-a-mobile-phone-news-photo/829576808#britains-prime-minister-tony-blair-looks-at-a-mobile-phone-camera-a-picture-id829576808 (accessed 17 December 2017).

Gentleman, D. (1987). *A Special Relationship*. London: Faber and Faber.

Gervasuti Foundation. (2011). *Pavilion of Iraq: Biennale Arte 2011*. Rome: Gangemi.

Ghazoul, R. (2017a). My work. http://www.rababghazoul.com/about (accessed 21 December 2017).

Ghazoul, R. (2017b). Small Medium Large | ManWoman Child. http://www.rababghazoul.com/work#/new-gallery-1/ (accessed 21 December 2017).

Gillan, K., Pickerill, J. and Webster, F. (2008). *Anti-War Activism: New Media and Protest in the Information Age*. Houndmills: Palgrave Macmillan.

Giovanni, J. (2010). Baghdad revisited: the Chalabi family's return from exile. *The Telegraph* (29 September). http://www.telegraph.co.uk/news/worldnews/middleeast/iraq/7906412/Baghdad-revisited-the-Chalabi-familys-return-from-exile.html (accessed 15 December 2017).

Goede, M. de (2014). Pre-emption contested: suspect spaces and preventability in the July 7 inquest. *Political Geography* 39: 48–57.

Goldenberg, S. (2003). Suddenly, the Iraq War is very real. *The Guardian* (19 March). https://www.theguardian.com/theguardian/2013/mar/19/iraq-war-saddam-bush-starts-2003 (accessed 1 December 2017).

Gough, P. (2010). *A Terrible Beauty: British Artists in the First World War*. Bristol: Sansom and Company.

GOV.UK. (2017). Iraq and Afghanistan Memorial unveiled in London. https://www.gov.uk/government/news/iraq-and-afghanistan-memorial-unveiled-in-london (accessed 12 April 2018).

Graham, S., ed. (2004). *Cities, War and Terrorism: Towards an Urban Geopolitics*. Oxford: Blackwell.

Graham, S. (2010). *Cities Under Siege: The New Military Urbanism*. London: Verso.

Graham, S. (2012). When life itself is war: on the urbanization of military and security doctrine. *International Journal of Urban and Regional Research* 36 (1): 136–155.

Gratton, P. (2014). *Speculative Realism: Problems and Prospects*. London: Bloomsbury.

Greater London Authority. (2018). The Invisible Enemy Should Not Exist by Michael Rakowitz. https://www.london.gov.uk/what-we-do/arts-and-culture/current-culture-projects/fourth-plinth-trafalgar-square/invisible-enemy-should-not-exist-michael-rakowitz (accessed 12 April 2018).

Gregory, D. (2004a). *The Colonial Present: Afghanistan, Palestine, Iraq*. Oxford: Wiley-Blackwell.

Gregory, D. (2004b). The angel of Iraq. *Environment and Planning D: Society and Space* 22 (3): 317–324.

Gregory, D. (2006a). Vanishing points. In: *Violent Geographies: Fear, Terror and Political Violence* (eds. Gregory, D. and Pred. A.), 205–236. London: Routledge.

Gregory, D. (2006b). The black flag: Guantánamo Bay and the space of exception. *Geografisker Annaler: Series B, Human Geography* 88 (4) 405–427.

Gregory, D. (2010a). War and peace. *Transactions of the Institute of British Geographers* 35 (2): 154–186.

Gregory, D. (2010b). Seeing red: Baghdad and the event-ful city. *Political Geography* 29 (5): 266–279.

Gregory, D. (2011). The everywhere war. *The Geographical Journal* 177 (3): 238–250.

Gregory, D., and Pred, A., eds. (2006). *Violent Geographies: Fear, Terror and Political Violence*. London: Routledge.

Gribble, R., Wessely, S., Klein, S., Alexander, D.A., Dandeker, C. and Fear, N.T. (2012). The UK armed forces: public support for the troops but not for the missions? In: *British Social Attitudes: The 29th Report* (eds. A. Park, E. Clery, J. Curtice, M. Phillips and D. Utting), 138–155. London: NatCen Social Research.

Gribble, R., Wessely, S., Klein, S., Alexander, D.A., Dandeker, C. and Fear, N.T. (2015). British public opinion after a decade of war: attitudes to Iraq and Afghanistan. *Politics* 35 (2): 128–150.

Grosz, E. (2008). *Chaos, Territory, Art: Deleuze and the Framing of the Earth*. New York: Columbia University Press.

Grosz, E. (2011). *Becoming Undone: Darwinian Reflections on Life, Politics and Art*. London: Duke University Press.

Hammilakis, Y. (2009). The 'War on Terror' and the military-archaeology complex: Iraq, ethics and neo-colonialism. *Archaeologies* 5 (1): 39–65.

Hammer, M. (2013). *Francis Bacon*. London: Phaidon.

Hanieh, A. (2015). A brief history of ISIS. *Jacobin* (12 March). https://www.jacobinmag.com/2015/12/isis-syria-iraq-war-al-qaeda-arab-spring/ accessed 15 December 2017.

Harvey, D. (2003). *The New Imperialism*. Oxford: Oxford University Press.

Haraway, D. (2016). *Staying with the Trouble: Making Kin in the Chthulucene*. Durham, NC: Duke University Press.

Harris, G. (2016). Portrait of Iraq's Picasso. *Financial Times* (4 November). https://www.ft.com/content/b1259018-99e5-11e6-8f9b-70e3cabccfae (accessed 12 July 2017).

Hashem, S. (2007a). *Freedom*. Exhibition catalogue.

Hashem, S. (2007b). Iraqi artists worldwide. http://www.incia.co.uk/31293.html (accessed 4 January 2018).

Hashem, S. (2013). *In Conflict: A Reflection on the Constant War in Iraq*. PDF document to accompany exhibition at the New Art Exchange, Nottingham.

Hawkins, H. (2010). 'The argument of the eye'? The cultural geographies of installation art. *Cultural Geographies* 17 (3): 321–340.

Heddaya, M. (2015). An outsourced vision? The trouble with Iraq's 'neocolonial' Venice Pavilion. *BlouinArtInfo* (2 April). http://www.blouinartinfo.com/news/story/1126765/an-outsourced-vision-the-trouble-with-iraqs-neocolonial (accessed 15 December 2017).

Herring, E., and Rangwala, G. (2006). *Iraq in Fragments: The Occupation and Its Legacy*. London: Hurst and Company.

Historic England. (2017). Royal Artillery Memorial. National Heritage List for England. Entry 231613. https://historicengland.org.uk/listing/the-list/list-entry/1231613 (accessed 17 December 2017).

Hodgkinson, T.W. (2016). Gerald Laing, the artist that made a war zone go pop. *Financial Times* (2 September). https://www.ft.com/content/d333aa80-70ed-11e6-a0c9-1365ce54b926 (accessed 17 December 2016).

Hörschelmann, K. (2008). Populating the landscapes of critical geopolitics – young people's responses to the war in Iraq (2003). *Political Geography* 27 (5): 587–609.

Hörschelmann, K., and Refaie, E. (2014). Transnational citizenship, dissent and the political geographies of youth. *Transactions of the Institute of British Geographers* 39 (3): 444–456.

Hudson, M. (2015). A twilight zone of the human psyche. In: *Tim Shaw* (ed. I. Khanna), 9–13. Bristol: Sansom and Company.

Hughes, B. (2013). Q&A with artist Martha Rosler. *Artnet News* (19 July). https://news.artnet.com/art-world/q-amp-a-with-artist-martha-rosler-49391 (accessed 1 August 2017).

Human Rights Watch. (1991). *Needless Deaths in the Gulf War: Civilian Casualties During the Air Campaign and Violations of the Laws of War*. New York: Human Rights Watch.

Hyndman, J. (2004). Mind the gap: bridging feminist and political geography through geopolitics. *Political Geography* 23 (3): 307–322.

Hyndman, J. (2007). Feminist geopolitics revisited: body counts in Iraq. *The Professional Geographer* 59 (1): 35–46.

Imperial War Museum. (2003). Art Commissions Committee of the Imperial War Museum awards research commission to Steve McQueen in response to the war in Iraq. Press notice. Facsimile held in IWM Research Library Collection.

Imperial War Museums. (2017a). History of IWM. http://www.iwm.org.uk/corporate/about-IWM (accessed 17 December 2017).

Imperial War Museums. (2017b). *Annual Report and Account 2016–2017* London: HMSO.

Imperial War Museums. (2017c). Photo Op. http://www.iwm.org.uk/collections/item/object/42971 (accessed 17 December 2017).

Ingram, A. (2009). Art and the geopolitical: remapping security at *Green Zone/Red Zone*. In: *Spaces of Security and Insecurity: Geographies of the War on Terror*. (eds. A. Ingram and K. Dodds), 257–277. Farnham: Ashgate.

Ingram, A. (2012). Bringing war home: from Baghdad 5 March 2007 to London 9 September 2010. *Political Geography* 31: 61–63.

Ingram, A. (2013). *Geographies of War | Iraq Revisited*. Exhibition catalogue. London: UCL.

Ingram, A. (2016). Rethinking art and geopolitics through aesthetics: artist responses to the Iraq war. *Transactions of the Institute of British Geographers* 41 (1): 1–13.

Ingram, A. (2017). Art, geopolitics and metapolitics at Tate Galleries London. *Geopolitics* 22 (3): 719–739.

Ingram, A., Forsyth, I. and Gauld, N. (2016). Beyond geopower: earthly and anthropic geopolitics in *The Great Game* by War Boutique. *Cultural Geographies* 23 (4): 635–652.

Intelligence and Security Committee (2003). *Iraqi Weapons of Mass Destruction – Intelligence and Assessments*. Norwich: The Stationery Office.

Ipsos MORI (2003). Iraq: the last pre-war polls. https://www.ipsos.com/ipsos-mori/en-uk/iraq-last-pre-war-polls (accessed 18 January 2019).

Iraq Body Count. (2018). Home page. https://www.iraqbodycount.org/ (accessed 11 April 2018).

Iraq Inquiry. (2016). Sir John Chilcot's public statement, 6 July 2016. https://webarchive.
nationalarchives.gov.uk/20171123124608/http://www.iraqinquiry.org.uk/the-inquiry/
sir-john-chilcots-public-statement/ (accessed 18 January 2019).

Ivakhiv, A. (2014). Beatnik brothers? Between Graham Harman and the Deleuzo-
Whiteheadian axis. *Parrhesia* 19: 65–78.

Jenkings, K.N., Megoran, N., Woodward, R. and Bos, D. (2012). Wootton Bassett and the
political spaces of remembrance and mourning. *Area* 44 (3): 356–363.

Jabra, J.I. (1961). *Art in Iraq To-day*. London: Embassy of the Republic of Iraq.

Jones, J. (2013). The Tony Blair 'selfie' Photo Op will have a place in history. *The Guardian*
(15 October). https://www.theguardian.com/artanddesign/2013/oct/15/tony-blair-selfie-
photo-op-imperial-war-museum (17 December 2017).

Jones, L. (2015). Tweet. (18 August). https://twitter.com/drleejones/status/63353687
9267332096 (accessed 17 December 2017).

Jordan, D. (2015). Image and after-image. In: *Tim Shaw* (ed. I. Khanna), 15–19. Bristol:
Sansom and Company.

Joronen, M., and Häkli, J. (2016). Politicizing ontology. *Progress in Human Geography* 4
(1): 561–579.

Kaiser, R. (2012). Reassembling the event: Estonia's 'Bronze Night'. *Environment and
Planning D: Society and Space* 30 (6): 1046–1063.

Kaplan, C. (2013). Sensing distance: the time and space of contemporary war. *Social Text
Online*. 17 June. https://socialtextjournal.org/periscope_article/sensing-distance-the-time-
and-space-of-contemporary-war/ (accessed 18 January 2019).

Karp, I., and Lavine S.D. eds. (1991). *Exhibiting Cultures: The Poetics and Politics of Museum
Display*. London: Smithsonian Institution Press.

Kavanagh, J. (2015). Brian Haw: the parliament square protestor. http://blog.museumoflondon.
org.uk/brian-haw-parliament-square-protestor/ (accessed 15 January 2018).

Keegan, J. (2005). *The Iraq War: The 21-Day Conflict and its Aftermath*. London: Pimlico.

Kennard, P. (1990). *Images for the End of the Century*. London: Journeyman Press.

Kennard, P. (2000). *Dispatches from an Unofficial War Artist*. London: Lund Humphries.

Kennard, P. (2015). *Unofficial War Artist*. London: Imperial War Museum.

kennardphillipps. (2011). Photo Op brought to life for Blair's reappearance at Iraq Inquiry.
https://vimeo.com/19105433 (accessed 17 December 2017).

Kennicott, P. (2004). A wretched new picture of America: photos from Iraq prison show
we are our own worst enemy. *Washington Post* 5 May, C01.

Khanna, I., and Shaw, T. (2015). A private and public ritual. In: *Tim Shaw* (ed. I. Khanna),
20–27. Bristol: Sansom and Company.

Kinney, A. (2006). *Hood*. London: Bloomsbury.

Kirsch, S., and Flint, C. (2011). Reconstruction and the worlds that war makes. In:
Reconstructing Conflict: Integrating War and Post-War Geographies (eds. S. Kirsch and C.
Flint), 3–28. Farnham: Ashgate.

Kitchin, R., and Dodge M (2007). Rethinking maps. *Progress in Human Geography* 31 (3):
331–344.

Klanten R., Bieber A., Alonzo P. and Krohn S., eds. (2011). *Art and Agenda: Political Art
and Activism*. Berlin: Gestalten.

Klein, N. (2007). *The Shock Doctrine: The Rise of Disaster Capitalism*. London: Allen Lane.

Klinke, I. (2013). Chronopolitics: a conceptual matrix. *Progress in Human Geography* 37
(5): 673–690.

Kluijver, R. (2010). *Borders: Contemporary Middle East Art and Discourse.* The Hague: Gemak.

Koopman, S. (2011). Alter-geopolitics: other securities are happening. *Geoforum* 42 (3): 274–284.

Kriebel, S. (2014). *Revolutionary Beauty: The Radical Photomontages of John Heartfield.* Berkeley: University of California Press.

Larsen, M. (1996). *The Conquest of Assyria: Excavations in an Antique Land.* London: Routledge.

Last, A. (2015). Fruit of the cyclone: undoing geopolitics through geopolitics. *Geoforum* 64: 56–64.

Last, A. (2017). We are the world? Anthropocene cultural production between geopoetics and geopolitics. *Theory, Culture and Society* 34 (2–3): 147–168.

Latour, B. (1987). *Science in Action: How to Follow Scientists and Engineers Through Society.* Cambridge, MA: Harvard University Press.

Latour, B. (1988). *The Pasteurization of France.* Cambridge, MA: Harvard University Press.

Laub, D. (1992). An event without a witness: truth, testimony and survival. In: *Testimony: Crises of Witnessing in Literature, Psychoanalysis and History* (eds. S. Felman and D. Laub), 75–92. London: Routledge.

Ledwidge, F. (2011). *Losing Small Wars: British Military Failure in Iraq and Afghanistan.* London: Yale University Press.

Legg, S. (2011). Assemblage/apparatus: using Deleuze and Foucault. *Area* 43 (2): 128–133.

Leighton House Museum. (2006). *Tears of the Ancient City: New Paintings by Suad al-Attar.* Exhibition catalogue. London: Leighton House Museum.

Leppard, D. (2006). Iraq terror backlash in UK 'for years'. *The Times* (2 April). https://www.thetimes.co.uk/article/iraq-terror-backlash-in-uk-for-years-k6qf2gxs7wt (accessed 13 December 2017).

Lewis, S., and Maslin, M. (2015). Defining the anthropocene. *Nature.* 519: 171–180.

Leys, R. (2010). *Trauma: A Genealogy.* Chicago, IL: University of Chicago Press.

Leys, R. (2011). The turn to affect: a critique. *Critical Inquiry* 37 (3): 434–472.

Lotz, C. (2009). Representation or sensation? A critique of Deleuze's philosophy of painting. *Symposium: Canadian Journal for Continental Philosophy* 13 (1): 59–73.

Loyd, J. (2009a). War is not healthy for children and other living things. *Environment and Planning D: Society and Space* 27 (3): 403–424.

Loyd, J. (2009b). 'A microscopic insurgent': militarization, health, and critical geogrpahies of violence. *Annals of the Association of American Geographers* 99 (5): 863–873.

Lundborg, T. (2009). The *becoming* of the 'event': a Deleuzian approach to understanding the production of social and political 'events'. *Theory & Event* 12 (1). DOI:10.1353/tae.0.042.

MacGregor, N. (2004). In the shadow of Babylon. *The Guardian* (14 June). https://www.theguardian.com/artanddesign/2004/jun/14/heritage.iraq (accessed 3 January 2018).

MacGregor, N. (2006a). Preface. In: *Word Into Art: Artists of the Modern Middle East* (ed. V. Porter), 9. London: British Museum.

MacGregor, N. (2006b). Spreading the word. *The Spectator* (20 May), pp.50–51.

Makiya, K. (as Al-Khalil, S.). (1989). *Republic of Fear: The Politics of Modern Iraq.* Berkeley: Regents of the University of California.

Makiya, K. (as Al-Khalil, S.). (1991). *The Monument: Art, Vulgarity and Responsibility in Iraq.* London: André Deutsche.

Makiya, K. (1993). *Cruelty and Silence: War, Tyranny, Uprising and the Arab World.* London: W.W. Norton.

Malallah, H. (2009). Sophisticated ways in the destruction of an ancient city. *Qui Parle* 17 (2): 103–108.

Malallah, H. (2010). Statement. (10 June). http://hanaa-malallah.com/words/statement. html (accessed 21 December 2017).

Martin, D. (2013). Ex-CIA officer on the strike that could have averted Iraq War. *CBS News* (19 March). http://www.cbsnews.com/news/ex-cia-officer-on-the-strike-that-could-have-averted-iraq-war/ (accessed 19 May 2017).

Massaro, V., and Williams, J. (2013). Feminist geopolitics. *Geography Compass* 7 (8): 567–577.

Massey, D., Bond, S. and Featherstone, D. (2009). The possibilities of a politics of place beyond place. *Scottish Geographical Journal* 125 (3–4): 401–420.

Massumi, B. (2002a). *Parables for the Virtual: Movement, Affect, Sensation.* Durham, NC: Duke University Press.

Massumi, B., ed. (2002b). *A Shock to Thought: Expression After Deleuze and Guattari.* London: Routledge.

Maunsell, F.R. (1897). The Mesopotamian petroleum field. *The Geographical Journal* 9 (5): 528–532.

Mbembe, A. (2001). *On the Postcolony.* Berkeley: University of California Press.

McBride, P. (2016). *The Chatter of the Visible: Montage and Narrative in Weimar Germany.* Ann Arbor: University of Michigan Press.

McQueen, S. (2010). *Queen and Country.* London: British Council.

McLagan, M., and McKee, Y., eds. (2012). *Sensible Politics: The Visual Culture of Nongovernmental Activism.* New York: Zone Books.

Mignolo, W.D. (2009). Epistemic disobedience, independent thought and decolonial freedom. *Theory, Culture & Society* 26 (7–8): 159–181.

Miller, T. (2015). Museums, ecology, citizenship. In: *The International Handbook of Museum Studies: Museum Theory* (eds. A. Witcomb and K. Message), 139–156. Oxford: Wiley-Blackwell.

Minioudaki, K. (2017). Sakar Sleman: Circles of suffering, remembrance and hope. In: *Archaic: the Pavilion of Iraq: 57th International Art Exhibition La Biennale di Venezia* (ed. S. Cernuschi), 76–79. Milan: Mousse.

Ministry of Defence (2017). *Iraq and Afghanistan Memorial 1990-2015.* Crown Copyright. https://assets.publishing.service.gov.uk/government/uploads/system/uploads/ attachment_data/file/596713/20170303_IAM_brochure_FINAL.pdf (accessed 18 January 2019).

Mitchell, T. (2013). *Carbon Democracy: Political Power in the Age of Oil.* London: Verso Books.

Mirzoeff, N. (2017). *The Appearance of Black Lives Matter.* [NAME]. https:// namepublications.org/item/2017/the-appearance-of-black-lives-matter/ (accessed 6 December 2017).

MORI. (2003). Iraq, the last pre-war polls. https://www.ipsos.com/ipsos-mori/en-uk/iraq-last-pre-war-polls (accessed 17 December 2017).

Morgan-Cox, R. (2016). *Gerald Laing 1936-2011. A Retrospective.* London: The Fine Art Society.

Morrissey, S. (2005). *Richard Wilson*. London: Tate.

Morton, M.Q. (2006). *In the Heart of the Desert*. Aylesford: Green Mountain.

Morton, T. (2007). Mark Wallinger. *Frieze* 107 (5 May). https://frieze.com/article/mark-wallinger-0 (accessed 17 December 2017).

Mosaic Rooms. (2012). TFL removes 'Iraq: How, Where, for Whom?' posters. http://mosaicrooms.org/blog/censored/ (accessed 17 December 2017).

Murphy, S. (2012). *The Art Kettle*. Winchester: Zero Books.

Muttitt, G. (2011). *Fuel on the Fire: Oil and Politics in Iraq*. London: The Bodley Head.

National Army Museum. (2016). *Consolidated Financial Statements for Year Ending 31st March 2016*. London: National Army Museum.

National Galleries Scotland. (2017a). Arthur Melville. https://www.nationalgalleries.org/art-and-artists/features/arthur-melville (accessed 6 December 2017).

National Galleries Scotland. (2017b). Peter Kennard and Cat Picton Phillipps. Photo Op. https://www.nationalgalleries.org/art-and-artists/101735/photo-op (accessed 17 December 2017).

National Portrait Gallery. (2017). Tony Blair ('Photo Op'). https://www.npg.org.uk/collections/search/portrait/mw232840/Tony-Blair-Photo-Op (accessed 17 December 2017).

Oliver, M. (2005). Accused soldier 'following orders'. *The Guardian* (19 January). https://www.theguardian.com/uk/2005/jan/19/military.world (accessed 17 December 2017).

O'Sullivan. S. (2005). *Art Encounters Deleuze and Guattari: Thought Beyond Representation*. London: Palgrave Macmillan.

Ó Tuathail, G. (1996). *Critical Geopolitics: The Politics of Writing Global Space*. London: Routledge.

Oxford English Dictionary (2018). Event. 11.3a. www.oed.com (accessed 14 September 2018).

Panigia, D. (2009). *The Political Life of Sensation*. Durham, NC: Duke University Press.

Papoulias, C., and Callard, F. (2010). Biology's gift: the turn to affect. *Body & Society* 16 (1): 29–56.

Park Gallery. (2017). Hanaa Malallah. Web page. http://theparkgallery.com/artist/hanaa-malallah/ (accessed 21 December 2017).

Patton, P. (1997). The world seen from within: Deleuze and the philosophy of events. *Theory & Event* 1 (1). DOI:10.1353/tae.1991.0006.

Patton, P. (2006). The event of colonisation. In: *Deleuze and the Contemporary World* (eds. I. Buchanan and A. Parr), 108–124. Edinburgh: Edinburgh University Press.

Pax, S. (2003). *The Baghdad Blog*. London: Atlantic Books.

Perks, S., Wilson, R. and Garmiany, A. (2012). Interview with Richard Wilson (transcript). https://web.archive.org/web/20121011141253/http://www.creativetourist.com/transcript/richard-wilson-sculptor (accessed 6 December 2017).

Petrešen-Bachelet, N. (2015). *Decolonising Museums*. L'Internationale Online. http://www.internationaleonline.org/media/files/decolonisingmuseums_pdf-final.pdf (accessed 15 December 2017).

Phillips, Richard. (2009). Common ground? Anti-imperialism in UK anti-war movements. In: *Spaces of Security and Insecurity: Geographies of the War on Terror* (eds. A. Ingram and K. Dodds), 239–256. Farnham: Ashgate.

Phillips, Ruth. (2007). Exhibiting Africa after modernism: globalization, pluralism and the persistent paradigms of art and artifact. In: *Museums After Modernism: Strategies of Engagement* (eds. G. Pollock and J. Zemans), 80–103. Oxford: Blackwell.

Philo, C. (2017). Less than human geographies. *Political Geography* 40: 256–258.

Pilar Blanco, M. del, and Peeren, E., eds. (2013). Introduction: conceptualizing spectralities. In: *The Spectralities Reader: Ghosts and Haunting in Contemporary Cultural Theory*, 1–27. London: Bloomsbury.

Pile, S. (2010). Emotions and affect in recent human geography. *Transactions of the Institute of British Geographers* 35 (1): 5–10.

Platt, S. Noyes. (2011). *Art and Politics Now: Cultural Activism in a Time of Crisis*. New York: Midmarch Arts Press.

Pollock, G. (2013a). *After-Effects | After-Images: Trauma and Aesthetic Transformation in the Virtual Feminist Museum* Manchester: Manchester University Press.

Pollock, G. (2013b). New encounters: arts, cultures, concepts. In: *Visual Politics of Psychoanalysis: Art and the Image in Post-Traumatic Cultures* (ed. G. Pollock), xiii–xvii. London: IB Tauris.

Pollock, G. (2013c). Editor's introduction. In: *Visual Politics of Psychoanalysis: Art and the Image in Post-Traumatic Cultures* (ed. G. Pollock), 1–22. London: IB Tauris.

Pollock, G., and Zemans, J., eds. (2007). *Museums After Modernism: Strategies of Engagement*. Oxford: Blackwell.

Porter, V. (2006). *Word into Art: Artists of the Modern Middle East*. London: British Museum Press.

Povinelli, E. (2011). *Economies of Abandonment: Social Belonging and Endurance in Late Liberalism*. Durham NC: Duke University Press.

Preziosi, D. and Farago, C.J. (2004). What are museums for? In: *Grasping the World: The Idea of the Museum* (eds. D. Preziosi and C.J. Farago), 1–9. Aldershot: Ashgate.

Proctor, R., and Schiebinger, L., eds. (2008). *Agnotology: The Making and Unmaking of Ignorance*. Stanford, CA: Stanford University Press.

Protevi, J. (2009). *Political Affect: Connecting the Social and the Somatic*. Minneapolis: Minnesota University Press.

Protevi, J. (2013). *Life, Earth, War: Deleuze and the Sciences*. Minneapolis: University of Minnesota Press.

Pyne, S.J. (2012). *Fire: Nature and Culture*. London: Reaktion Books.

Pyne, S.J. (2009). The human geography of fire: a research agenda. *Progress in Human Geography* 33 (4): 443–46.

Quijano, A. (2007). Coloniality and modernity/rationality. *Cultural Studies* 21 (2–3): 168–178.

Radvansky, G.A., and Zacks, J.M. (2014). *Event Cognition*. Oxford: Oxford University Press.

Rakowitz, M. (2007). The invisible enemy should not exist. Web page. http://www.michaelrakowitz.com/the-invisible-enemy-should-not-exist/ (accessed 17 December 2017).

Rancière, J. (2002). The aesthetic revolution and its outcomes: emplotments of autonomy and heteronomy. *New Left Review* 14: 133–151.

Rancière, J. (2006). *The Politics of Aesthetics*. London: Continuum.

Rancière, J. (2009a). *Aesthetics and its Discontents*. Cambridge: Polity.

Rancière, J. (2009b). Contemporary art and the politics of aesthetics. In: *Communities of Sense: Rethinking Aesthetics and Politics* (ed. B. Hinderliter), 31–50. Durham, NC: Duke University Press.

Rancière, J. (2010). *Dissensus: On Politics and Aesthetics*. London: Continuum.

Rappert, B. (2012). *How to Look Good in a War: Justifying and Challenging State Violence*. London: Pluto Press.

Re-Piano Project (2010). Instruments of Connection. Project flyer. In author's possession.

Ricks, T. (2006). *Fiasco: The American Military Adventure in Iraq*. London: Penguin.

Riverbend. (2006). *Baghdad Burning: Girl Blog from Iraq*. London: Marion Boyars.

Robbins, D. (2006). Introduction. In: *Tears of the Ancient City: New Paintings by Suad al-Attar*, 5–8. Exhibition catalogue. London: Leighton House Museum.

Robinson, A. (2014). Alain Badiou: the event. *Ceasefire* 15 December, https://ceasefiremagazine.co.uk/alain-badiou-event/ (accessed 23 October 2018).

Rosler, M. (2004). *Decoys and Disruptions: Selected Writings 1975–2001*. Cambridge, MA: MIT Press and International Center of Photography.

Ross, A. (2011). Geographies of war and the putative peace. *Political Geography* 30 (4): 197–199.

Routledge, P. (2005). Reflections on the G8 protests: an interview with General Unrest of the Clandestine Insurgent Rebel Clown Army. *ACME: An International E-Journal for Critical Geographies* 3 (2): 112–120.

Ruya Foundation. (2013). Foreword. In: *Welcome to Iraq: the Pavilion of Iraq at the 55th International Art Exhibition La Biennale di Venezia* (eds. T. Chalabi and J. Watkins). Baghdad: Ruya Foundation for Contemporary Culture in Iraq.

Ruya Foundation. (2017). Venice conversations: place, memorial and memory in Sakar Sleman's work. (26 April). https://ruyafoundation.org/en/2017/04/4850/ (accessed 15 December 2017).

Sahakian, R. (2012). It is what it is. *Jadaliyya*. (22 February). http://www.middleeastdigest.com/pages/index/4461/it-is-what-it-is (accessed 2 April 2018).

Sahakian, R. (2016). On the closing of Sada for Iraqi art. (6 April). http://www.warscapes.com/blog/closing-sada-iraqi-art (accessed 15 December 2017).

Said, E. (1978). *Orientalism*. London: Penguin.

Said, E. (1994). *Culture and Imperialism*. London: Vintage.

Saleh, Z. (1966). *Britain and Mesopotamia (Iraq) 1600 to 1914: A Study in British Foreign Affairs*. Baghdad: Al-Ma'aref.

Sanger, D.E., and Burns, J.F. (2003). Bush declares start of Iraq war; missile said to be aimed at Hussein. *The New York Times*. (20 March). http://www.nytimes.com/2003/03/20/international/worldspecial/bush-declares-start-of-iraq-war-missile-said-to.html (accessed 1 December 2017).

Sands, P. (2005). *Lawless World: The Making and Breaking of Global Rules*. London: Allen Lane.

Satia, P. (2008). *Spies in Arabia: The Great War and the Cultural Foundations of Britain's Covert Empire in the Middle East*. Oxford: Oxford University Press.

Sawyer, P. (2007). Army museum shows art linking 7/7 to Iraq. *Evening Standard* (2 July). https://www.standard.co.uk/arts/army-museum-shows-art-linking-77-to-iraq-war-6594573.html (accessed 17 December 2017).

Scarry, E. (1985). *The Body in Pain: The Making and Unmaking of the World*. New York: Oxford University Press.

Schatzki, T.R., Knorr-Cetina, K. and Savigny E., eds. (2001). *The Practice Turn in Contemporary Theory*. London: Routledge.

Schatzki, T.R. (1996). *Social Practices: A Wittgensteinian Approach to Human Activity and the Social*. Cambridge: Cambridge University Press.

Schatzki, T.R. (2001). Introduction: practice theory. In: *The Practice Turn in Contemporary Theory* (eds. T.R. Schatzki, K. Knorr-Cetina and E. Savigny), 10–23.

Schuster, J. (2014). How to write the disaster. *Minnesota Review* 83 (1): 163–171.

Scott, A. (1990). *Domination and the Arts of Resistance: Hidden Transcripts*. London: Yale University Press.

Scranton, R. (2015). *Learning to Die in the Anthropocene: Reflections on the End of a Civilization*. San Francisco, CA: City Lights.

Secor, A. (2007). An unrecognizable condition has arrived. In: *Violent Geographies: Fear, Terror and Political Violence* (eds. D. Gregory and A. Pred), 37–53. London: Routledge.

Selim, R. (2018). Ark Reimagined. http://culturunners.com/projects/art-re-imagined (accessed 4 April 2018).

Sewell, W. (1996). Historical events as transformations of structures: inventing revolution at the Bastille. *Theory and Society* 25 (6): 841–881.

Shabout, N. (2007). Presenting Iraqi visual culture. In: *Dafatir: Contemporary Iraqi Book Art* (ed. Shabout, N.), 10–25. Denton TX: University of North Texas Press.

Shabout, N. (2009). "A dream we call Baghdad". In: *Modernism and Iraq* (eds. Z. Bahrani and N. Shabout), 23–40. New York: Miriam and Ira D. Wallach Art Gallery, Columbia University.

Shapiro, M.J. (2012). *Studies in Trans-Disciplinary Method: After the Aesthetic Turn*. London: Routledge.

Sharpe, C. (2016). *In the Wake: On Blackness and Being*. Durham, NC: Duke University Press.

Shaw, I.G.R. (2012). Towards an eventral geography. *Progress in Human Geography* 36 (5): 613–627.

Sladen, M. (2007). Introduction. In: *Memorial to the Iraq War*, unpaginated. London: Institute of Contemporary Arts.

Sluglett, P. (2007). *Britain in Iraq: Contriving King and Country*. London: I.B. Tauris.

Smith, S. (2016). Befriended by a king, arrested, then forced to fight… Artist Dia Azzawi on the destruction of his beloved Iraq. *The Telegraph* (17 October). http://www.telegraph.co.uk/art/artists/befriended-by-a-king-arrested-then-forced-to-fight-artist-dia-az/ (accessed 12 July 2017).

Sontag, S. (2003). *Regarding the Pain of Others*. London: Penguin.

Stallabrass, J. (2013). *Memory of Fire: Images of War and the War of Images*. Brighton: Photoworks.

Steele, J. (2009). *Defeat: Why They Lost Iraq*. London: IB Tauris.

Stewart, R. (2006). *Occupational Hazards*. London: Picador.

Steyerl, H. (2013). *Is the Museum a Battlefield*. https://vimeo.com/76011774 (accessed 3 January 2018).

Stoler, A.L. (2008). Imperial debris: reflections on ruins and ruination. *Cultural Anthropology* 23 (2): 191–219.

Stop the War Coalition. (2011). *Stop the War: A Graphic History*. London: Francis Boutle and Stop the War Coalition.

Sumartojo, S. (2013). The Fourth Plinth: creating and contesting national identity in Trafalgar Square 2005–2010. *Cultural Geographies* 20 (1): 67–81.

Sylvester, C. (2009). *Art/Museums: International Relations Where We Least Expect It*. London: Paradigm.

Sylvester, C. (2013). *War as Experience: Contributions From International Relations and Feminist Analysis*. Abingdon: Routledge.

Szyłak, A. (2017). Sherko Abbas: music of the Bush era. In: *Archaic: the Pavilion of Iraq: 57th International Art Exhibition La Biennale di Venezia* (ed. Cernuschi, S.), 40–43. Milan: Mousse.

Tate. (2007). New work: Mark Wallinger. http://www.tate.org.uk/context-comment/video/new-work-mark-wallinger (accessed 17 December 2017).

Taylor, P.J. (1982). A materialist framework for political geography. *Transactions of the Institute of British Geographers* 7 (1): 15–34.

Taylor, P.J. (1993). *Political Geography: World-Economy, Nation-State, and Locality*. Harlow: Longman.

Thomas, O. (2015). The Iraq Inquiries: publicity, secrecy and liberal war. In: *Liberal Wars: Anglo-American Strategy, Ideology and Practice* (ed. A. Cromartie), 128–149, Abingdon: Routledge.

Thompson, N., and Independent Curators International. (2009). *Experimental Geography: Radical Approaches to Landscape, Cartography, and Urbanism*. New York: Melville House.

Toal, G. (2008). Book Forum: Derek Gregory's *The Colonial Present* (Oxford, Blackwell, 2004). *Political Geography* 27 (3): 339–343.

Toal, G., and O'Loughlin, J. (2017). Why did MH17 crash? Blame attribution, television news and public opinion in Southeastern Ukraine, Crimea and the de facto states of Abkhazia, South Ossetia and Transnistria. *Geopolitics* 23 (1): 1–35.

Tripp, C. (2007). *A History of Iraq*. Cambridge: Cambridge University Press.

Tuhiwai Smith, L. (1999). *Decolonizing Methodologies: Research and Indigenous Peoples*. London: Zed Books.

Tversky, B., and Zacks J.M. (2013). Event perception. In: *The Oxford Handbook of Cognitive Psychology* (ed. D. Reisberg). Oxford: Oxford University Press. DOI: 10.1093/oxfordhb/9780195376746.013.0006.

Valentine, G. (2007). Theorising and researching intersectionality: a challenge for feminist geography. *Professional Geographer* 59 (1): 10–21.

Vazquez, R., and Mignolo, W. (2013). Decolonial aesthesis: colonial wounds/decolonial healings. *Social Text Online* (15 July). https://socialtextjournal.org/periscope_article/decolonial-aesthesis-colonial-woundsdecolonial-healings/ (accessed 3 December 2017).

Verso. (2016). After Chilcot: a reading list on Iraq and the 'War on Terror'. https://www.versobooks.com/blogs/1627-after-chilcot-a-reading-list-on-iraq-and-the-war-on-terror (accessed 17 December 2016).

Visser, I. (2011). Trauma theory and postcolonial literary studies. *Journal of Postcolonial Writing* 47 (3): 270–282.

Visser, I. (2015). Decolonizing trauma theory: retrospect and prospects. *Humanities* 4 (2): 250–265.

Visser, R. (2009). Proto-political conceptions of Iraq in late Ottoman times. *International Journal of Contemporary Iraqi Studies* 2 (1): 143–154.

Visser, R. (2013). Dammit, it is NOT unraveling: an historian's rebuke to misrepresentations of Sykes-Picot. *Iraq and Gulf Analysis*. https://gulfanalysis.wordpress.com/2013/12/30/dammit-it-is-not-unravelling-an-historians-rebuke-to-misrepresentations-of-sykes-picot/ (accessed 2 April 2018).

Wallerstein, I. (1974). *The Modern World System: Vol. 1, Capitalist Agriculture and the Origins of the European World Economy in the Sixteenth Century*. London: Academic Press.

Walters, W. (2014). Drone strikes, *dingpolitik* and beyond: furthering the debate on materiality and security. *Security Dialogue* 45 (2): 101–118.

Wetherell, M. (2012). *Affect and Emotion: A New Social Science Understanding*. London: Sage.

Whitehead, M., Jones, M., Jones, R., Woods, M., Dixon, D. and Hannah, M., eds. (2014). *An Introduction to Political Geography: Space, Place and Politics*. London: Routledge.

Willett, J. (1978). *The New Sobriety: Art and Politics in the Weimar Period 1917–33*. London: Thames and Hudson.

Williams, Alison. (2014). Disrupting air power: performativity and the unsettling of geopolitical frames through artworks. *Political Geography* 42: 12–22.

Williams, James. (2009). If not here, then where? On the location and individuation of events in Badiou and Deleuze. *Deleuze Studies* 3 (1): 97–123.

Williams, Raymond. (1977). *Marxism and Literature*. Oxford: Oxford University Press.

Wodiczko, K. (2011). War, Conflict and Art. https://www.youtube.com/watch?v=iGVfGTu3kkM. Accessed 15 December 2017.

Wolfe, P. (2006). Settler colonialism and the elimination of the native. *Journal of Genocide Research* 8 (4): 387–409.

Woodward, R., and Jenkings, N. (2012). Soldiers' photographic representations of participation in armed conflict. In: *Representations of Peace and Conflict* (eds. S. Gibson and S. Mollan), 105–119. Basingstoke: Palgrave Macmillan.

Woodward, R., and Jenkings, N. (2016). The uses of memoir in military research. In: *the Routledge Companion to Military Research Methods* (eds. A. Williams, N. Jenkings, R. Woodward and M. Rech), 71–83. London: Routledge.

YouGov. (2015). Memories of Iraq: did we ever support the war? https://yougov.co.uk/news/2015/06/03/remembering-iraq/ (accessed 17 December 2017).

Yusoff, K. (2015). Geologic subjects: nonhuman origins, geomorphic aesthetics and the art of becoming inhuman. *Cultural Geographies* 22 (3).: 383–407.

Yusoff, K. (2016). Anthropogenesis: origins and endings in the Anthropocene. *Theory and Society* 33 (2) 3–28.

Yusoff, K., Grosz, E., Clark, N., Saldhana, A. and Nash, C. (2012). Geopower: a panel on Elizabeth Grosz's *Chaos, Territory, Art:* Deleuze and the framing of the earth. *Environment and Planning D: Society and Space* 370 (6): 971–988.

Žižek, S. (2014). *Event: Philosophy in Transit*. London: Penguin.

Index

Geopolitics and the Event: Rethinking Britain's Iraq War Through Art, First Edition. Alan Ingram.
© 2019 Royal Geographical Society (with the Institute of British Geographers).
Published 2019 by John Wiley & Sons Ltd.